Starter Coops

For Your Chickens' First Home

By Wendy Bedwell-Wilson

Project Team
Lead Editor: Jennifer Calvert
Senior Editor: Amy Deputato
Senior Editor: Jarelle S. Stein
Art Director: Cindy Kassebaum
Book Project Specialist: Karen Julian
Production Coordinator: Leah Rosalez
Indexer: Melody Englund

i-5 PUBLISHING, LLC™
Chief Executive Officer: Mark Harris
Chief Financial Officer: Nicole Fabian
Chief Content Officer: June Kikuchi
Chief Digital Officer: Jennifer Black
Chief Marketing Officer: Beth Freeman Reynolds
General Manager, i-5 Press: Christopher Reggio
Art Director, i-5 Press: Mary Ann Kahn
Senior Editor, i-5 Press: Amy Deputato
Production Director: Laurie Panaggio
Production Manager: Jessica Jaensch

Front Cover Photography: Furtwangl/Flickr. **Back Cover Photography:** (top) Furtwangl/Flickr, (bottom) Ruthdaniel3444/Flickr

Library of Congress Cataloging-in-Publication Data
Bedwell-Wilson, Wendy.
Starter coops : for your chickens' first home / by Wendy Bedwell-Wilson.
p. cm.
Includes bibliographical references and index.
ISBN 978-1-935484-77-6
1. Chickens--Housing. I. Title.
SF494.5.B43 2012
636.5--dc23

2012011721

i-5 Publishing, LLC™
3 Burroughs, Irvine, CA 92618
www.facebook.com/i5press
www.i5publishing.com

Printed and bound in China
15 16 17 3 5 7 9 8 6 4

Dedication

Greenhorn chicken keepers, this book is dedicated to you. I hope it helps you create a cozy home-sweet-home for your hens and sets you up for success with your new hobby!

Acknowledgments

This book really exists because of one person: my farmer husband, Ryan. The main man in charge of our livestock and land, he has been my go-to guy for all things chicken. *Starter Coops* is steeped in his hands-on experience in raising and keeping chickens and building henhouses and chicken coops for them. His love for the hobby has truly inspired this book.

Besides Farmer Ryan, my second go-to resource for chicken keeping is Gail Damerow and her books *Storey's Guide to Raising Chickens* and *The Chicken Health Handbook*. These indispensable reference books are packed full of everything you need to know to raise hens—and then some.

I'd also like to offer a big thanks to our local feed store owners Mike and Stan Jackson from Central Feed and Supply in Sutherlin, Oregon, for their on-the-ground chicken advice, as well as Roger Sipe, editor of *Chickens* magazine, who planted the seed for this book when he asked me to write a regular column, "Coop Corner," in his bimonthly title.

Of course, *Starter Coops*' behind-the-scenes crew deserves a nod too! Special thanks to lead editor Jen Calvert and copy editors Amy Deputato and Jarelle Stein, who turned my manuscript into an easy-to-understand guide for new chicken keepers; editor-in-chief Andrew DePrisco, who believed in the idea; and the production team who made this book a reality.

Finally, I'd like to recognize Jerry, Larry, and Lance, our roosters. Those bad-ass chickens kept our ladies clucking and laying safely and happily. Thanks for being such great guardians and alarm clocks, boys!

Contents

Introduction

Welcome to the Henhouse!

Throughout history, humans have raised chickens for both practical and pleasurable purposes. Families and farmers across the globe kept flocks of birds for eggs, meat, and by-products such as feathers and manure. Extra birds became tradable goods to barter at markets. In short, chickens were a nice commodity to have around.

I'll never forget my first encounter with domesticated chickens. While visiting my family's farm in rural Iowa in the 1970s, my cousins and I were tasked with gathering eggs for breakfast. As a six-year-old California native raised in the Silicon Valley suburbs, I had never gazed upon a live chicken, walked into a henhouse, or reached into a nesting box to harvest freshly laid eggs. But once we stepped foot inside the sprawling chicken coop, I marveled at the little birds scurrying and scratching through the soil for seeds, seemingly unperturbed by the young human interlopers. I watched, fascinated, as the cackling roosters steered their ladies toward wriggling earthworms

Chickens are some of the most rewarding creatures to keep, not just because of the eggs they provide but also for their entertaining personalities.

and bright-green grass shoots. I wondered why in the world they didn't fly away. After we gingerly transferred the still-warm eggs into our woven-willow basket and delivered them to my great aunt, I declared that I, too, would have chickens one day.

That day finally came not too long ago. In 2008, my husband and I, relatively inexperienced at the whole farming thing, moved to an 80-acre piece of Pacific Northwest paradise in southern Oregon. Our first order of business: order some fuzzy peeps and raise a chicken coop. It seemed simple at the time—but boy, did we have a lot to learn. First was the matter of housing the baby chicks and keeping them warm and cozy while pinfeathers replaced their downy, dusty fluff. Then, when the birds were mature enough to graduate from the brooder to the newly built henhouse, we discovered (through trial and error) precisely what chickens needed to be happy and healthy as well as to stay safe and secure from resident raccoons, skunks, and coyotes.

Admittedly, the learning curve was steep. But thanks to the helpful feed-store staff, some weathered farmer friends, lots of reading and research, and a few close calls, our entire six-hen-and-one-rooster flock of Rhode Island Reds and Plymouth Barred Rocks survived and thrived through the first season.

Today our flock has grown to twenty-two hens, two roosters, and two Peking ducks. (Heritage breeds are next on our chicken wish list.) Needless to say, we've had to make room for our ever-expanding poultry brood. As we've deepened our knowledge of chicken keeping and poultry housing, we've relocated our chicken yard, made changes to the henhouse configuration, and experimented with nest-box designs. We even have plans to build a chicken tractor so the ladies can fertilize our nutrient-sapped pastureland.

Few things are cuter than a box full of peeps, but these fuzzy little guys will require plenty of attention and effort from you.

While chickens can, of course, provide meat, many keepers feel squeamish about picking off their well-kept flock. Eggs, however, are a natural (and nutritious) by-product of chicken keeping that everyone can appreciate.

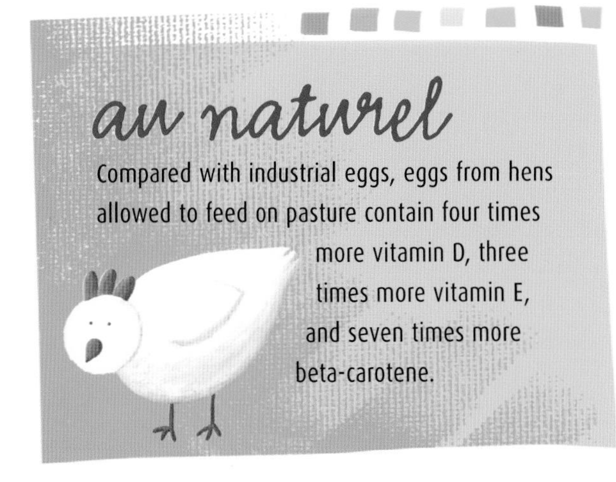

au naturel

Compared with industrial eggs, eggs from hens allowed to feed on pasture contain four times more vitamin D, three times more vitamin E, and seven times more beta-carotene.

As with all things farming, we still have a lot to learn—and always will! But from my coop to yours, I'd like to humbly share what my husband and I (and our chickens) have learned about chicken housing. That's what this book is all about.

In 2008, we weren't the only newbie chicken keepers on the block. Countless others across the country (and the world) began to see the value of keeping their own chickens. Whether prompted by the global economic downturn, a trendy return to homesteading, or a growing awareness of factory-farmed birds' often-deplorable living conditions, people began buying and raising hens like mad. They realized they could inexpensively raise their own organic farm-fresh eggs and meat with no hormones or dangerous chemicals. They could make use of the chickens' nitrogen-rich manure in their compost piles and vegetable gardens. And they could house the hens in top-notch living quarters rather than in crowded cages. The chicken-keeping resurgence had begun!

In historic terms, chicken keeping is nothing new. Humans have shared their lives with domesticated birds for thousands of years—even as far back as 6000 BC. A Chinese archaeo-zoologist named Chao Ben-shuh uncovered a large quantity of chicken bones from an archaeological site called the P'ei-li-kang village settlements of T'zu Shan in eastern Asia. Those bones were dated to more than 7,000 years ago, providing evidence of some of the world's first domesticated chickens, though it's not known whether these early predecessors were truly related to today's fowl.

Modern-day chickens likely descended from birds of Indian origin for the purposes of cockfighting in Asia, Africa, and Europe. Believe it or not, early poultry keepers had little interest in chicken meat

Some unsavory characters still enjoy watching cockfights today. Responsible chicken keepers work hard to avoid these kinds of painful brawls.

or eggs. The sport of rooster sparring spread the birds westward, and chickens migrated into eastern Europe along the Mediterranean coasts by around 3000 BC.

By 1400 BC, ancient Egyptians recognized the value of chickens for sustenance. Needing to fuel the huge labor force required to build the now-famed pyramids, they developed hatching ovens to mass-produce chickens for their eggs and meat. Historians report that each facility produced 15,000,000 to 20,000,000 chicks per season. The operators lived on the premises and tended the warming ovens. Because they had no thermometers to help regulate the temperature inside, workers relied on their senses to recognize when the fires needed attention. Large-scale peep housing and production have certainly come a long way!

SQUAWK BOX

"While keeping backyard chickens was common 150 years ago, the advent of factory farming and inexpensive store-bought eggs in the 1950s lead to a decline in its popularity. Recently, however, there has been a resurgence of interest in keeping one's own chickens, both for the pleasure of fresh eggs and for the entertainment pet chickens provide."

Hilary Stern, DVM

As time passed, chickens continued to flock across Europe. The first pictures of chickens on the Continent appeared on Corinthian pottery from the seventh century BC. Early Greek poets and authors, including Cratinus and Aristophanes, wrote about the birds in their works from the mid-fifth century BC. From 1200 to 200 BC, in ancient Greece and Rome, chickens were reportedly used as oracles in a practice known as *alectryomancy*, which involves prophesying based on the movements and behaviors of chickens.

After the turn of the century, the Roman agriculture author Lucius Junius Moderatus Columella, who lived from around AD 4 to 70, wrote about chicken husbandry in his twelve-volume treatise *De Re Rustica*. In his eighth book, he identified several cockfighting breeds—Tanagrian, Rhodic, Chalkidic, and Median—and wrote that Roman chickens, a cross between docile native hens and Greek roosters, were better for farming purposes. An ideal flock, he wrote, consisted of 200 birds, which could be guarded and tended by one person on the lookout for predators. He advised early chicken keepers to build three-room chicken coops that faced southeast, adjacent to the kitchen, and that contained a hearth to keep the ladies cozy and warm. He even recommended that early farmers provide dry dust or ash for the chickens to bathe in.

Polynesian seafarers then helped spread chickens across the globe. In the twelfth century AD, chickens reached Easter Island, where they were the only domestic animal and were housed in extremely solid chicken coops built from stone. As time passed, they reached every continent, supplying residents with eggs and meat.

From the late nineteenth century forward, Americans took a liking to chickens. The American Poultry Association, a breed registry, formed in 1873 to document the various chicken breeds raised in the United States. By the 1930s in the United States, chicken farms had begun cropping up both in the country and in nearby cities to feed America's growing population. Initially, large flocks of free-range birds spent much of their time outdoors and lived in a variety of chicken coops.

As the population's penchant for poultry increased over the following decades, the free-range farming style transitioned into factory production methods. Factory farms were common by the 1950s. Scientists discovered that chickens would lay more eggs and get fatter faster if they confined the birds indoors and kept the lights on at night.

Though production increased, quality decreased. Large numbers of cooped-up birds required large doses of antibiotics to stave off infections, and poultry farmers had to blunt the birds' beaks to prevent them

As you can see, factory farms rob chickens of the pasture-grazing life they're genetically engineered to lead.

Bird Brain

According to the USDA, in the United States in 2010:

- **The value of all egg production was $6.52 billion (up 6 percent from $6.17 billion in 2009).**
- **Egg production totaled 91.4 billion eggs (up 1 percent from 90.5 billion eggs produced in 2009).**
- **The average US resident gobbled down 246.2 eggs.**
- **The top ten egg-producing states were Iowa, Ohio, Pennsylvania, Indiana, California, Texas, Minnesota, Michigan, Nebraska, and Florida.**

from pecking one another. Rather than follow their natural foraging patterns of scratching and pecking for seeds, bugs, and greens in the great outdoors, the chickens were banished to windowless pens and fed a controlled diet of mush.

This treatment, which many perceived as inhumane, eventually came to light. As a result, consumers began to demand chickens that were raised in open pens or pastures rather than in cages. They wanted birds that were treated humanely. Some of these outspoken advocates for fair chicken treatment—people like you and me—even decided to raise their own.

If all of the reports in newspapers and magazines are any indication, backyard chicken keeping is a trend that's here to stay. People have discovered the joys and fringe benefits of owning chickens, so more and more are trying their hand at taking care of their own flocks. These pioneering chicken keepers are in cities and suburbs as well as in the country. At any given moment, more than 10 billion chickens worldwide are clucking happily, laying close to 700 billion eggs each year.

All those chickens need a place to live—chicken coops to call their own—and this book aims to help you design the best coop for your flock. As you flip through these pages, you'll learn what type of shelter

- giving you a place where you can collect the fringe benefits of hen keeping—eggs, manure, and so on; and
- being a place for must-have items and accessories, such as a station for food and water.

> **SQUAWK BOX**
>
> *"All chickens have the same basic requirements to stay healthy: a good-quality diet, a clean environment, and protection from the elements and predators."*
>
> *Hilary Stern, DVM*

To sum up, chicken coops—regardless of size or shape—must be designed to be practical and functional for both the owner and the birds. Your coop should provide adequate shelter, be sized appropriately for your flock, give you easy access to eggs and litter, and be outfitted with the right accessories. Seems pretty straightforward, right? Let's take a closer look at these different functions and the reasons they're vital to your flock's health, happiness, and well-being.

Give Me Shelter

All animals—feathered or otherwise—need food, water, and shelter to survive. One of the most critical functions a chicken coop serves is as a safe, enclosed space for the birds that protects them from predators and the elements. In addition to the varieties I mentioned above, shelter options for your ladies include A-frame structures, a covered stoop or porch, and even bushes and shrubbery—anything that shields them from danger.

A fortress-strong chicken coop is critical to a flock's health and well-being because, well, chickens are on the wrong end of the food chain. All kinds of predators—from hawks and owls to cats, raccoons, and

No matter what your coop looks like, it should be strong, sturdy, and safe for the chickens that occupy it.

coyotes—lurk high and low in the city, suburbs, and country. A well-designed and sound chicken coop with plenty of shelter options (including the henhouse) can protect the birds from danger. Simply put, a safe enclosure gives the birds a place to scurry to and hide should danger approach.

Henhouses and shelters also protect the birds from bad weather conditions. When temperatures climb into the 80s and 90s, the enclosure offers your flock shade. When temperatures dip, it gives your birds a place to huddle together, generating body heat and keeping away from chilling drafts. And when it rains or snows and the wind blows, the shelter provides a warm, dry place for your flock to weather the storm.

Home, Sweet Home

A chicken coop also serves as your flock's home base. It's where the birds live. It's where they go about their day-to-day routines, such as foraging for food, taking dust baths (see page 33), and snoozing. If they're laying hens, they deposit their eggs in the coop. In short, the coop is where your chickens feel safe and comfortable.

Add all the chicken requisites to your coop, and your hens will return to it each evening.

Chickens are so content in their henhouse that they return to their roosts each and every night at dusk, like clockwork, on their own. My husband and I learned that fact when we raised our first flock of birds. New to the hobby and unaware of the birds' natural homing instinct, we were reluctant to let them out of their wired-in shelter to roam freely in our fenced half-acre yard, afraid that at the end of the day we would have a difficult time getting them back into their coop. Finally, we decided to let them loose and see what happened. To our delight, they returned to their chicken house without us (or our canine chicken herder, Pete) having to (gently) convince them to do so.

Chickens like predictability. They're creatures of habit. By providing a home base and a sound structure for them to retire to every night, you'll minimize their stress and help them feel secure—which ultimately means you'll have happy chickens and yummy eggs.

Fringe Benefits

If you're like most backyard chicken keepers, you're likely raising the birds so you can harvest your own eggs and/or meat and collect the birds' droppings for your compost heap. A well-designed chicken coop gives you easy access to the goodies inside as well as a way to efficiently clean and sanitize the henhouse.

It's truly a hobby farmer's delight to gather fresh eggs for breakfast—but imagine if you had to hunt for them in the chicken yard as you would Easter eggs! Nesting boxes in your chicken coop give the birds a convenient place to lay their eggs in an easy-for-you-to-gather place where they won't be trampled by

If you don't provide nesting boxes for your laying hens, you might come out to find filthy and broken eggs.

scampering chickens, broken or pecked by broody hens, or stolen by egg-loving thieves, such as skunks. Clean, litter-lined nesting boxes also keep the eggs clean and dry.

While we're talking about by-products, there's nothing like aged chicken manure to heat up the soil in your vegetable garden. You could scoop and shovel the nitrogen-dense litter from the henhouse into a wheelbarrow and haul it to your compost pile. However, by incorporating easy-access designs into the coop, such as a sliding tray beneath the birds' roosts for droppings, you can easily gather and utilize the manure without having to get too messy in the process.

Amenities, Please

Finally, as your girls' home base, a chicken coop also houses all the accessories the birds will need to thrive. They will need a dedicated station for food and water with easy-to-clean vessels for feed, fresh water, grit, oyster shells, and any other supplement or tasty ingestible. For laying hens, nesting boxes are a must-have, as are perches for roosting safely off the ground at night. The ladies will also benefit from a dust-bath box they can use to keep their feathers clean and parasite-free. These things, which we'll discuss in detail in the next section, are the little frills that keep chickens happy and healthy. Continue reading and you'll learn how to incorporate these concepts into your design.

Chickens and gardens are mutually beneficial. You can use your chickens' manure to fertilize your garden, and you can plant chicken-friendly flora to treat your girls.

Coop Features and Decor

When it comes to making your ladies feel right at home, it's all about the niceties. Think about it: the features that make your own house a home include cozy comforts such as plush carpeting, an overstuffed couch and big-screen television, central heating and air conditioning, and granite countertops, right? It's a similar story with chickens. They thrive in a coop with clean and fluffy bedding, private nesting boxes, adequate ventilation, comfortable temperatures, and clean and sanitized food and water dishes.

With some architectural planning and interior coop-design savvy, it's easy to ensure your chickens' henhouse and yard provide not only the special touches that make chickens comfortable but also the essential elements that they need to flourish. Here's what you should consider before you break ground on your ladies' new digs.

In addition to a hen's breed and age, her egg output will be affected by environmental factors, including temperature and light. Hens lay best when the temperature is between 45 and 80 degrees Fahrenheit. When the weather is cooler or warmer than that, production slows. Egg laying slows or stops in the winter, too, when the number of daylight hours falls below fourteen. To extend "daylight" hours and prolong the laying season, some keepers turn on a full-spectrum light in their chickens' abode.

Your chicken coop can be more than just functional. Add some personal touches to give it some personality.

Architectural Elements

First things first. Coop comforts begin with a discussion about the chicken coop's structure and its conveniences. As you'll learn in chapter 2, your chickens' yard and henhouse can be designed in myriad ways depending on your particular needs. Regardless of its design, however, the coop will need to incorporate some specific architectural features. It should

- be free from drafts and able to maintain a comfortable ambient temperature,
- offer plenty of space for the birds to stretch their wings and wander,
- provide adequate ventilation,
- offer protection against predators—both airborne and terrestrial—and
- be accessible so that you can clean and collect eggs.

Let's take a closer look at why these elements are important and how you can incorporate them into your chicken coop.

Temperature, Draft Control

When designing your henhouse and yard, you need to make temperature and draft control top priorities. Chickens prefer ambient temperatures between 70 and 75 degrees Fahrenheit—room temperature for us humans. Because the birds live outside, where temperatures vary depending on where you live and what season it is, you'll need to consider architectural features that ensure the birds stay comfortable.

In warmer weather, the little ladies stay cool by drinking water; lounging under shade trees, shrubbery, or awnings; or chilling in their henhouse. Some building features that can help your girls keep cool include

- strategically placed shades outside the henhouse, such as awnings, tarps, or A-frames;
- screened windows that can be opened and closed during warm weather;
- wall and roof insulation, particularly on the south and west sides, to keep the hot temperatures outside the shelter;
- outdoor misters that spray water and lower the air temperature; and
- electricity to operate fans that will circulate air.

In colder weather, the birds stay warm by eating more feed to generate more heat, fluffing up their feathers to trap a layer of warm air near their skin, and huddling together in the coop to share body heat with their fellow fowl. The trouble is that cold drafts and breezes blowing through the shelter and yard can wick away their heat. Wind can cause warm air close to the birds' bodies to be replaced by cooler air. During wintertime, this

Chickens that live in warmer climates need places to cool off, such as under shade trees and shrubs or awnings.

When temperatures drop, you'll need to keep your hens warm. A heat lamp is a great resource.

heat transfer can cause a bird to chill. To make sure your girls stay warm, particularly in colder climates, consider these architectural features:

- a draft-free henhouse, particularly at chicken level (meaning on the ground or in their perches);
- double-wall construction;
- wall and roof insulation to keep the warm air inside the shelter, along with a vapor barrier to keep moisture to a minimum;
- straw bales stacked against the henhouse's north wall;
- south-facing windows that allow winter sun in; and
- electricity to run heat lamps placed strategically over perches.

Bottom line: No matter where you live, make sure your coop will provide a comfortable, temperate environment for your ladies. Doing so will keep them healthy, help them grow, and give them what they need to produce top-quality eggs.

Give Them Space

As semidomesticated barnyard critters, chickens need the right amount of habitable space to live in. Although some factory-farming operations have shown how little space chickens need, the birds will be healthier and happier if they have lots of room to spread their wings and scurry around the yard. Sure, they'll sometimes huddle together in little feather piles under eves and in their henhouse, but they appreciate the room to roam—particularly any hens lower in the pecking order that are feeling bullied.

When thinking about your chicken coop's construction, keep in mind that the space—including the yard and the henhouse—should be sized appropriately for the type and number of birds in your flock, as well as for their age. So precisely how much space do your ladies need? Take a look.

A small coop like this one is perfect for hobby farmers who want to keep just a couple of hens.

Chicks: Peeps and young birds aged one day to one week need less than 1 square foot per bird. As the chicks grow, their space needs will grow with them. Because they're young and unable to defend themselves, they should be kept in a brooder or in a confined shelter, such as a fully fenced-in coop.

Pullets and young birds: Chickens aged one week through twenty weeks or so can move into an open pen (or one with an accessible yard) that allows 1 to 3 square feet per bird, or a confined shelter that allows 2 to 7 square feet per bird. Heavier breeds, such as Jersey Giants, will require more space than lighter breeds, such as Rhode Island Reds; tiny Bantams will require even less.

Adult hens and roosters: By the time the birds are twenty-one weeks, they'll be fully grown. If they're living in an open pen, they'll need 3 to 4 square feet per bird; in a confined shelter, they'll need 7 to 10 square feet per bird. Again, heavier breeds will need more room than lighter ones.

Birds are adaptable and can live in tighter quarters than what I've described here. But if you start to see behavior issues such as hens fighting or picking at each other's feathers, poor growth, or reproductive problems, your girls may need a little more legroom.

Fresh-Air Friendly

Chicken coops don't produce the most, well, appealing of scents. A poorly ventilated henhouse can be downright malodorous. Those odors result from the chickens' natural respiratory and digestive processes. When chickens breathe, they emit moisture and carbon dioxide; when they digest food, they emit methane. Add to that the ammonia, hydrogen sulfide, and carbon monoxide that develop from the chickens' droppings—not to mention flying dust and debris—and you have a stinky cocktail of

This screened window provides ventilation and a nice view for perching chickens but can be closed off at night by lowering the wooden hatch to ward off the cold and nocturnal predators.

airborne gas and particles that can affect the hens' respiratory passages and overall health. That's why ventilation is so critical.

Essentially, ventilation circulates the air inside the henhouse, allowing moist, contaminated air to exit while bringing in fresh, clean air. Your goal is to facilitate that airflow while reaching a humidity level of between 40 and 60 percent; any more than that and the moisture in the coop can actually be considered a health hazard because a host of viruses and bacteria thrive in moist air.

Talk to someone at your local cooperative extension or animal control office to find out which predators plague your area.

When designing your henhouse, plan to cut some ventilation openings near the top of the structure (out of the chicken zone) to give warm, moist air a place to escape and cool, drier air a place to enter. This is especially important during warm summer days because high temperatures could affect your birds' health, comfort, and productivity. Most experts recommend carving decent-size openings—windows that are 1 foot by 3 feet, for instance—along the south and north walls, covering them with chicken wire, and constructing hinged drop-down covers that allow the windows to open or close as the weather dictates.

Predator Proofed

Poor little chickens. It seems like everything is out to get them. Whether you live in the city, the suburbs, or the country, your hens are a prime target for a variety of airborne and terrestrial predators; these predators threaten to take the hens' eggs or the birds themselves. Prime suspects include

Nesting and bedding materials—no matter their form—are soft and absorbent, lightweight and easy to handle, affordable, and not treated with toxic chemicals. Available options include:

- **Straw.** Offered at farm and feed stores, straw is a relatively inexpensive material to use in the chicken coop. It's also compostable, making it a great choice for gardeners.
- **Pine shavings.** Shavings are also available from farm and feed stores, as well as from pet stores. Pine shavings cost a bit more, but they smell nice, they're absorbent, and they're also compostable.
- **Shredded newspaper.** This is an excellent option for those who want to use recycled, compostable material, and it's inexpensive (if you have a newspaper subscription and shred the paper yourself). However, newspaper is only moderately absorbent, so you'll need to replace it often.
- **Dried lawn clippings.** If you don't spray your lawn or pasture with chemical pesticides or herbicides, you can use well-dried lawn and grass clippings you've collected after mowing. You're growing it anyway—why not use it?

Straw and pine shavings make excellent bedding choices, and you can find both inexpensively at your local farm-supply store.

au naturel

Farmers have used this trick for centuries. Rather than clean out the henhouse manure weekly (or more often, depending on the brood), they would simply toss a layer of fresh bedding on top of the old. Come spring, they would shovel the layers of aged litter into their fields as fertilizer.

You can do something similar in your henhouse. After you've done your annual spring-cleaning, lay down a 4- to 6-inch layer of clean bedding. When the surface gets packed or matted, break it up using a shovel and add enough fresh bedding on top to absorb the moisture. Keep adding fresh litter as needed; you'll be surprised by how quickly the volume reduces once the pile starts to naturally compost. Rather than become smelly and filthy, properly managed litter gradually ferments, raising the temperature in the henhouse and keeping the ladies warm during the cooler months. Another bonus: flies are less of a problem because the manure attracts natural fly predators and parasites.

If you plan to use this method in your henhouse, make certain the dwelling is well ventilated and the manure has the right amount of moisture—not too much, not too little. You'll know the litter is too damp if it begins to smell of ammonia. When you're ready to harvest the fertilizer, shovel it into your wheelbarrow, take it out to where you need it, and begin the process again in your henhouse.

Chickens love to perch, and they'll do it on any flat surface they can get their feet on.

Try out different types of bedding to see which works best for your particular situation. Whichever you choose, make sure you freshen it often to keep it dry and droppings-free. That way, you'll also keep the odor at bay and your ladies' respiratory systems healthy.

Perches and Roosts

Laying hens and roosters prefer to perch off the ground where they can grasp a branch, fluff their feathers around their feet, and cozy in with their flock. Perches are not mandatory design elements in a chicken coop, but they're a nice feature to add to the henhouse because they give your chickens a place to rest at night, exercise their toes, smooth down their nails, regulate their body temperatures while sleeping, and even play.

Chickens prefer rough, strong surfaces, especially hewn wood with rounded corners sized for them to grip comfortably with their toes—think tree branches thick enough to support their weight. In general, the perch for a regular-size chicken should be about 1½ to 2 inches wide; the perch for Bantams should be no less than 1 inch wide.

Some possible perch-making materials include

- untreated pine 2x2s with rounded tops;
- wooden dowels, about 2 inches in diameter;
- strong branches from your yard that have been stripped of their bark;
- old pieces of lumber that you've inspected for nails and splinters, sanded down, and ripped to the right length and width;
- ladders made from wood; and
- any other item that's strong enough to hold the birds and rough enough for them to grip without being so splintery as to injure their feet.

Avoid plastic pipe and metal pipe, both of which are too smooth for the chickens to grasp firmly.

The length of your chickens' perches will depend on your flock's size. Each bird will require about 6 to 10 linear inches of space, so if you have ten birds, plan for between 5 and 8 feet (60 and 96 inches) of perch. To give the birds enough space to comfortably roost, you should allow about 18 inches between the roost and the wall, or between parallel roosts. If your ladies live in tight quarters, you can create steps of roosts 12 inches apart vertically and horizontally so that the chickens can easily hop from lower to higher rungs. See page 79 for a useful perch project.

Hoppers

Two more important interior design elements are the feed station and the supplement hopper, where you provide your chicken with their daily rations and necessary supplements, such as grit or oyster shell. You can find hanging tubes, gravity feeders, plastic or metal bowls, crocks, troughs, and other containers at your local feed store. In general, the stainless-steel varieties last longer and are easier to sanitize than the plastic ones. You can also create your own feeder. For some time, we've used a 4-foot length of gutter flipped right-side up and leaned against the side of the henhouse for a feeder. It's long and low, easy to hose out, and light enough that the birds don't try to perch on it because it'll tip over.

Allow space for at least two feeders and a hopper—even if you have a small flock—to ensure that the weaker birds don't get bullied away. If you have roosters, furnish at least one feeding station per rooster. Each rooster will gather his hens around his feeder, which will reduce fighting. Put out enough feeders so that at least one-third of your flock can eat simultaneously. As a general rule, allow 4 inches of feeder for each mature chicken.

But where should the feeders and hopper go? If you put them inside the henhouse, the feed will stay dry, but the birds may spend a lot of time cooped up. If you put them outside the henhouse

Thick wooden dowels make perfect perches—just don't expect them to stay clean for long.

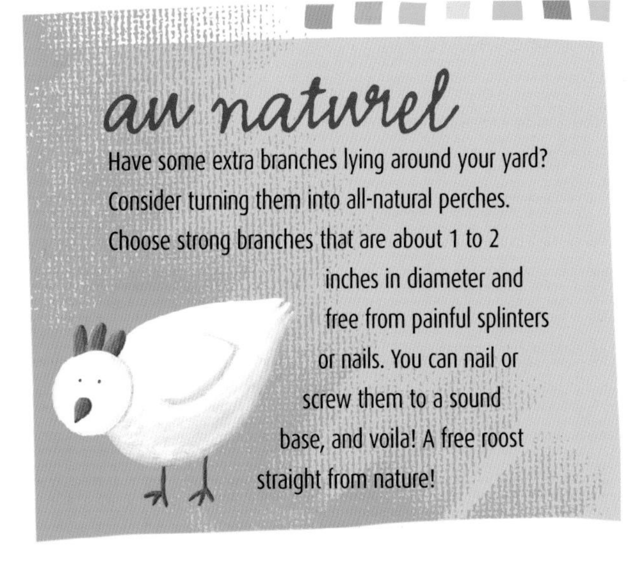

au naturel

Have some extra branches lying around your yard? Consider turning them into all-natural perches. Choose strong branches that are about 1 to 2 inches in diameter and free from painful splinters or nails. You can nail or screw them to a sound base, and voila! A free roost straight from nature!

Feeders come in all shapes and sizes. The important thing is that your feeder is large enough to accommodate all of your chickens.

under a weatherproof awning, the birds will get fresh air, but the food may attract unwanted visitors. During the dry season, our ladies dine al fresco. (To prevent critters from stealing the food at night, we put the food inside their locked henhouse.) During the rainy season, the food and supplements stay nice and dry inside. Inside or out, the food station's location should be easy for the flock to reach yet far enough from the perches that the birds don't inadvertently soil the food.

Store your feed in a clean, dry, rodent-proof container. Keep it in a cool area away from light, heat, and moisture.

Water Stations

Like the food and supplement stations, the birds' water stations should be centrally located and easy for the flock to reach. Water is critical to your birds' health, so they should have plenty of watering holes to visit in their chicken coop. The best waterers are

- leak-free, drip-free, and stable enough that they won't easily tip;
- large enough to supply the chickens with water all day;
- easy to clean and sanitize; and
- designed to prevent chicken-generated debris from soiling the water.

For a small flock, hand-filled and hand-carried bell-shape waterers work great. They come in a range of sizes and materials; galvanized waterers are generally preferred because, frankly, they last longer and weigh more than the plastic variety, which means they aren't easily knocked over. They can also be hung, which will keep the water cleaner longer. In a pinch, you could use buckets, plastic planter boxes, and even crocks and bowls—but these types of container should be cleaned out frequently because they'll get dirty very quickly.

Because larger flocks need more water throughout the day, piped-in water may be a good option. The system automatically refills the waterers, virtually ensuring that the chickens will have water at all times. However, these systems aren't without their challenges. In addition to the cost of running plumbing out to your coop, leaky or frozen pipes are a risk. Another option is to set up a gravity flow system in which water is stored in a tank and refills the dish when the water drops to a certain level.

No matter what system you use, you should provide enough waterers so that at least one-third of your birds can drink at the same time. Even if you have so few chickens that one drinking station appears to be adequate, provide two—spaced apart—to ensure that all chickens can get a drink without fighting or chasing away those lower in the pecking order.

Again, regardless of which kind of waterer you choose, make sure it's large enough and you have enough of them so that every hen can quench her thirst.

Dust-Bath Box

Chickens love to take dust baths. They fluff down into a divot filled with dry dirt, toss the dust up into the air, and roll around in the hole, flapping their feathers and kicking their legs. When they're done, they hop up and shake themselves off, leaving behind a puff of dust. This dirty-dancing routine, which is typically followed by a preening session, keeps their feathers clean and in good condition. It also helps a chicken rid its feathers of parasites.

If your ladies don't have a yard in which they can pick they're favorite spot, make sure your henhouse includes a box for dust baths.

If you plan to let your ladies roam in the yard, they'll likely find their own spot for dust baths in your garden. If you keep your chickens in a confined space, however—or the rainy season has swamped the free-rangers' favorite bathing holes—you'll need to provide a special place for them, such as a low-sided box or litter pan filled with dust. A bright light placed overhead will prevent them from treating it as a nesting box.

There's No Place Like Home

Chickens are hearty birds and can adapt to virtually any living and housing situation. To help your hens thrive, however, you should provide them with an ideal environment, which includes a secure chicken coop, comfortable temperatures, plenty of room, and all the comforts of home. In the next chapter, you'll learn more about the different types of coops for chickens.

Tour de Coops

If you've never been on a chicken-coop tour, I highly recommend taking one of these informative and fascinating tours, which have been growing in popularity as more and more people acquire their own backyard flocks. During these open-henhouse tours (conducted much like garden or home tours), owners welcome visitors into the chicken coops to meet the birds, see the setups, and gather ideas for their own hens' yards.

Tours can be found in various places throughout the United States. My first coop crawl, cleverly titled Tour de Coops, was through Lyons, Colorado—an up-and-coming town just outside of Boulder. My sister, her four young daughters, and I had the pleasure of visiting half a dozen homes, all of which had distinctly different chicken coops. One downtown homeowner created hers from an existing dog run and a small prefab henhouse from her local feed store. Another backyard flock keeper, who lived farther out in the rural Apple Valley area, built a large fully wired-in chicken yard with a ramshackle henhouse constructed from repurposed barn lumber; a colorful hand-painted sign above the coop wished everyone a sunny "Good Morning!" Still another fowl fancier—a rocket scientist by trade—erected in his suburban backyard a self-designed top-of-the-line fully insulated henhouse, complete with running water and electricity.

I spoke at length with the coop owners, took copious notes, and filled my camera's memory card with photos. With so many great ideas, I simply had to incorporate some of them into my own chicken-coop design. The lesson I learned from the tour: when it comes to a coop, just about anything goes as long as it provides for the hens' basic needs. The shapes, sizes, and configurations are limitless.

Almost anything can be used as a chicken coop—including a train. Live poultry trains were used by brokers in the United States in the 1920s to transport birds across the country. Each car could accommodate nearly 5,000 chickens and was equipped with a water tank, a grain container, and a room for a chicken keeper. The eggs laid were property of the train crew.

So Many Options...

Chickens' easy-going real-estate requirements are one reason they've been such popular barnyard critters for thousands of years. Most keepers meet those requirements by using one of four basic types of chicken coops. Which one you choose (and ultimately how you modify it for your specific needs and personalize it for your own liking) will depend on your own situation. In the sections that follow, I'll introduce you to the four different ways that poultry fanciers house their hens:

- free range, which is seen most often in rural areas
- in permanent dwellings, which is the most common housing method for small flocks
- in portable dwellings, which are used on farms and in family gardens with lots of land
- In specialized housing, which includes dwellings for show chickens and baby birds

Read on for more details—including the pros and cons—about these different sheltering methods as well as how the henhouses and designs differ from each other.

Free Range

Free range is a popular phrase these days. When printed on egg cartons and poultry packaging, free range alludes to the way that chickens are housed. It conjures up images of happy hens frolicking in a sunny green pasture, gobbling up bugs and nibbling on grass shoots, as opposed to unhappy hens confined to an indoor-only dwelling. These marketing claims, of course, may or may not be true—and that uncertainty is likely one reason you decided to raise your own chickens in the first place!

True free-range living means that the wild or semidomesticated birds have, well, *free range* of the land they live on. They roam unbound, foraging for food and nesting (and laying eggs) where they see fit. If they're wild, they may have access to a shelter of some kind or roost in a tree for safety; if they're part of a barnyard clan, they'll likely return to their roost each night—particularly if they're fed and watered there.

FREE RANGE

Pros

Chickens love chomping on bugs, slugs, spiders, and even mice. They'll circle your yard, hunting for protein- and calcium-rich snacks. If you have free-range ladies, you could eliminate the need for chemical pesticides and rodent and slug poisons. (It's not safe to use these chemicals, anyway, if your chickens are roaming free.)

Your chickens will rake and scrape the ground with their claws while they nibble seeds and green foliage shoots. This action efficiently controls weeds and spreads mulch around your garden.

Really, it's a hoot to watch and listen to the hens go about their hunt-and-peck business. If one hen finds a worm or tasty bug, the other ladies scramble over to her, trying to get their piece of the invertebrate. They caw and cackle to one another in their own chicken language and can be more entertaining than television.

Cons

Without the protection of a fence or shelter, your chickens could be dinner for a neighborhood cat, raccoon, or other carnivorous critter. Even if they can find safety perched in a tree or crouched in some shrubbery, free-range hens are easy targets for predators.

Similarly, hens that roam free may also be exposed to harsh weather, such as extreme heat or cold, which could threaten their health.

Depending on where you live, keeping free-roaming chickens might not be the most neighborly thing to do. Your hens may cross property lines and turn up the neighbors' gardens or leave messy droppings on their lawns or patios. Not everyone loves these feathered friends the way we do.

Free-range chickens run around during the day, but they still need a coop to return to at night.

A common problem with free-ranging chickens is that they can escape the boundaries of your property.

A pro/con list is all well and good, but the real question is: Is free ranging right for you and your charges? As idyllic as this type of chicken keeping may sound, in most cases you should—as a responsible chicken keeper—confine your birds to the safety of a chicken coop. However, you can do so and still reap the benefits of free-roaming birds by housing them in a mobile pen, which we'll discuss in a little bit (see page 41).

SQUAWK BOX

"Chickens are very adaptable, and no single best way exists to house them. Creative architectural construction may even be considered in building a 'designer' chicken house in order to enhance the backyard landscape."

David Frame, DVM

Permanent Dwellings

A permanent henhouse is the traditional method for housing homestead poultry and other small backyard flocks. Depending on the farm and the number of birds being kept, these built-to-last structures can vary in size from thousands of square feet to just a few dozen square feet.

Generally speaking, there are two types of permanent chicken coops: confined housing (those designs that keep the birds within the structure) and yarded housing (those designs that allow the hens access to a fenced yard). Let's look at the features of different permanent dwellings, and then examine each one's benefits and drawbacks.

Confined Housing

This large permanent coop resembles an aviary.

Confined housing provides living space for your ladies within an enclosed pen; the chickens are not let out of their coop into an open yard or pasture. Pens like this can be used for small backyard flocks living in tight urban or suburban yards, for raising broilers or breeders, when maintaining a flock during cold or wet weather, and even as a quarantine pen for birds that you're sequestering for some reason.

A coop like this typically resembles a large cage or aviary with a small henhouse inside. Within the tightly woven-wire or fenced enclosure, the birds have easy access to their food and water station, nest boxes and perches (if laying hens), and dust-bath area. Depending on the coop owner's preference,

CONFINED HOUSING

Pros

Because the birds are confined in one place, cleanup and care are easy, as is collecting compost from the coop's floor.

Being confined also guarantees that your chickens won't be ravaging your vegetable garden or leaving droppings on your back deck.

Confined housing works great if other pets, such as cats and dogs, share the backyard.

If you build the coop large enough, you can hang out with your feathered friends whenever you please.

Confined coops suit young chicken keepers as well, from those doing 4-H projects to those caring for the chickens as pets.

You may need to keep the birds contained for their own safety, especially if you have animals that roam the backyard.

Cons

Confined housing limits the number of birds you can keep. Chicken keeping can be addictive, particularly when you get into heritage breeds and the like, but you can only house so many in a confined space.

It will require more maintenance—scooping compost, adding fresh bedding, and so on—to keep the coop clean and sanitary.

The ladies may not be able to engage in natural behaviors, such as foraging for weeds and bugs. Some feel this hampers the hens' quality of life.

If the hens get into spats, the lowest chicken on the totem pole might be harassed and even injured. Like humans, hens have their own personalities; even if you provide enough space in the coop for everyone, squabbles may occur.

If a predator gets in, your birds are toast because they have nowhere to go. A friend of mine had six birds in a confined-housing-style coop. One day, their Dachshund slipped inside and slaughtered the flock. The poor chickens couldn't get away from dog.

If you never move a mobile coop, it might be considered a permanent structure, but it won't be as safe as one.

the enclosure could have an earthen floor, a wire floor over a droppings pit, or even a concrete or other easy-to-clean floor. A confined coop can even be kept indoors in a well-ventilated basement or barn—as long as you clean it regularly.

An important feature of a confined-housing-style coop is its size relative to your flock's size: it should give the birds enough room to exercise, scurry about, and spread their wings comfortably. It also should be sound and secure so that no predators can get in, and no birds can get out.

Is this type of housing right for you? Possibly. If space is tight in your backyard and you don't want to deal with chickens running amok, confined housing is a great option. This would also work well if you have space but it's not fenced for chickens. Or if your child wants to keep a chicken as a pet. Confined housing works well in many situations.

Yarded Housing

Unlike confined housing, where the birds are kept within a structure, yarded housing gives chickens access to an outdoor fenced yard as well as providing a permanent structure for sleeping, perching, egg laying, and protection. These coops give your ladies all the benefits of free ranging, while offering them more safety from predators and shelter from the elements. If you have the space on your property, this is an excellent way to keep your chickens.

Yarded housing essentially consists of a henhouse with all its niceties as well as a large, securely fenced open yard where the birds can run around and flap their wings to their little hearts' content. In some cases, the yard may be covered with netting or poultry wire to foil flying predators. The yard can include sun and wind protection, such as A-frames, awnings attached to the henhouse, and small range (lean-to) shelters—particularly useful if the yard has no shade trees. Elaborate chicken yards may even include hen-friendly foliage and trees to encourage them to eat their greens.

We keep our little ladies and roosters in a yarded-housing-style chicken coop. Their henhouse opens into a fenced 1,000-square-foot pasture, which we divided (so that we can rotate their living space) and planted with nutrient-rich clover and wheatgrass. Chickens love apples and the

If you have a fenced-in yard, "yarded housing" can mean simply adding a henhouse. Just make sure to check the perimeter for any vulnerabilities.

You could also construct a fence that isn't anchored into the ground so that you can move it and rotate the areas your chickens can access.

YARDED HOUSING

Pros

As with confined housing, yarded housing sequesters the ladies in one place. However, they have access to more open space where they can engage in free-range behavior, such as scratching for seed, eating greens, and hanging out with other hens.

When provided with hiding places and secure fencing around the coop, the chickens are (for the most part) safe from predators and foul weather.

With yarded housing, you can expand your flock more easily than with confined housing, as long as the henhouse is sized appropriately.

This type of housing works well if you have any chicken-unfriendly pets (these could include dogs and cats—though agile cats may be able to hop the fence and harass the birds anyway).

Cons

Honestly, fencing is expensive. It can cost a pretty penny to invest in posts, concrete, heavy-gauge chicken-safe fencing, and ties for a large poultry yard. You'd have to sell a lot of eggs to pay for the tools and materials.

Unless you cover the yard, the birds may be more vulnerable to predators from above. With no roof or overhead protection, chickens can be easily whisked away by raptors.

A flock of chickens can denude a pasture or expensive landscaping job in no time. If you want your girls to eat their greens, you'll need to replant regularly.

Because the birds are roaming free in the yard, their droppings will fall where they may—and that can create a muddy mess, particularly in wet climates. Yard rotation can help. This is when you rotate your ladies between a number of yards cross-fenced within your chicken coop, allowing the elements to break down the droppings.

bugs they attract, so we planted two apple trees in each yard for fruity treats and shade. An A-frame and an old table umbrella give the birds shade and protection from resident red-tailed hawks, while a small plastic swimming pool and water buckets give them cool water to drink. They love it!

Is it the right housing solution for your birds? If you have the space and funds to do it, definitely. You don't need to have acres and acres of land. Even a petite 10-by-10-foot space gives a small flock 100 square feet of freedom.

If your property is home to hawks, take precautionary measures.

Mobile Homes for Chickens

Another option for chicken keepers is a mobile home. Portable chicken coops and pens are easily moved around the yard or pasture, providing the birds with an enclosed space for safety and roosting as well as access to fresh greens—and giving you the benefit of having a larger area of land fertilized (a different version of yard rotation). Because these poultry RVs need to be light enough for one or two people to lift, they tend to be smaller than their permanent alternatives.

Your homemade chicken tractor can have everything this prefabricated one has: wheels, a handlebar, and a henhouse with easy access.

In general, you'll find two types of portable chicken coops: a chicken tractor, or a mobile coop that keeps the birds within the structure; and a fenced range, which limits the hens' grazing area to within a mobile fence. As we did with the permanent structures, let's check out these portable dwellings' features and weigh the pros and cons.

The Chicken Tractor

No, a chicken tractor (also called an ark) isn't a farming implement for your hens to drag around the pasture (though that would be nice, wouldn't it?). It's a confined floorless portable shelter that you can move daily or weekly to give the birds access to fresh nutritious greens and new ground to scratch. While the birds nibble and peck, they're aerating and fertilizing the ground beneath them; as you move the tractor across the garden or field over time, every part of it gets the chicken treatment. This method works well in large family gardens and in fields where you have enough pasture for the shelter to be moved frequently.

Chicken tractors are constructed of tight-mesh chicken wire and lumber, usually shaped like an A-frame, a rectangle, or a square topped with a roof—whatever lightweight (but not too lightweight!) solution works for your particular situation. Many times, chicken tractors have handles, skids, or

A simple ark might be an easier DIY task—just a few pieces of lumber and some wire and you have yourself a coop.

CHICKEN TRACTOR

Pros

The birds can live in a safe, contained space while spreading their manure across a pasture or garden area and aerating it with their claws.

They have access to fresh greens and bugs to eat—but only in areas where you let them graze. That means your prized roses will be safe from nibbling beaks.

You'll have less cleanup to do because the chickens will spread the droppings themselves as the you move the tractor.

Even though the chickens are mobile, you can still collect the eggs or grow the chickens as broilers.

With a chicken tractor, your chickens get the nutritional benefits of grazing and your grass gets pruned and deloused.

Cons

Because the ground is not always level, the pen may not sit flat, and that could mean an entry point for predators. Some bad guys could also burrow beneath the pen and nab unsuspecting hens.

To get the full benefit of the mobile pen, you need to move it regularly. How often depends on the size of your flock and the size of the pen itself. A small 4-by-6-foot pen with two birds, for instance, may need to be moved weekly; a larger 8-by-16-foot pen with a half dozen hens could stay in place for two weeks or longer.

If the chickens aren't used to being tractored, they may get bored. That can result in behavior challenges, such as spats or intermittent feather plucking.

Depending on where you live, tractoring may not be right for every season. If you live where it regularly snows in the winter, for instance, your girls could develop frostbite.

wheels to make moving them easier. Like the permanent structures, these units have perches (for laying hens), nest boxes, and food and water dishes. The structure's height doesn't matter too much; the width and depth will make the difference as to how many birds you can fit inside.

We have colleagues who tractor their broilers. They have a good-size hobby farm with about 20 acres of fertile pasture, and they've designed mobile pens using PVC pipe, tarps, and fine-mesh fencing. The pens are low-lying structures that are easy to move yet heavy enough to stay put on windy days. They're easy to build—and they're cheap.

Is a chicken tractor the right housing for your chickens? If you have the property to rotate the chickens and their pen, and a partner in crime who can help you move the tractor, then yes indeed. We're thinking about trying this out next spring.

Fenced Range

Just as yarded housing gives hens more room to roam than confined housing does, a fenced range gives tractored birds more room too. This type of mobile chicken coop features a movable pen, just like a chicken tractor, but it's surrounded by a movable fence, which gives the chickens more space to scurry around in. Imagine a dog kennel inside an exercise pen: the dog can retire to its own den, or it can play in the enclosed area. (See the pros and cons for both yarded housing and chicken tractors.)

This chicken is outside looking in on his fenced range. The fence isn't anchored, so it can be moved wherever it's needed.

This type of chicken coop does a great job at giving tractored chickens more room to roam, hunting and pecking for food, but it can be a challenge to uproot and relocate fencing every few weeks or so.

Special Housing

In the course of your chicken-keeping adventure, you may find yourself in need of some special housing. You'll need a brooder, or a coop for young chicks, at the beginning of your journey. As your chickens grow up, you may discover you have a rooster that needs to be periodically segregated from the crowd. Or flock dynamics (and drama) may require you to move a hen-pecked gal to her own digs. If one of your ladies gets sick or if a hen wants to hatch her own eggs, she may need to be sequestered from the group. All of these situations require special housing away from the flock's coop. Despite the variety of these needs, you'll commonly find two kinds of special housing:

- a brooder, which is a shelter for peeps and young chicks
- an isolation pen, which is a one-chicken shelter where a bird can have its own space

If peeps are in your future, you'll need to invest in a brooder or build one. Luckily, brooders are easy to set up.

Brooder Basics

A brooder is a self-contained peep shelter that holds everything the little chicks will need during their first few weeks of life, including bedding, food, water, and warmth. You can purchase commercial brooder boxes or you can fashion your own using a range of materials, from a rubber tote to a built-from-scratch plywood box. See pages 82 to 84 for a basic plan.

The size of your brooder will depend on the number of peeps you have (or plan to have). When they're young, the birds need about 1 square foot apiece; that means that if you have a half-dozen feather fluffs running around, you'll need a 3-by-2-foot space. Their space needs will increase as they grow—and it'll happen more quickly than you think.

Here are some additional features your brooder will need:

- **Ventilated lid.** To keep the peeps from hopping out of their home, your brooder will need a lid of some kind. A mesh screen allows for good ventilation, but it can release the heat that the birds need to thrive. A good alternative is a solid lid with several holes cut out and covered with mesh.
- **Solid ground.** As for the flooring, experts recommend a solid floor lined with absorbent litter, such as wood shavings or shredded newspaper, that's changed regularly. Mesh flooring is an option, but it has its disadvantages, including the nasty habit of cannibalism (peeps on mesh flooring are unable to peck at the ground, which is a natural chicken behavior, so they may peck at one another's toes).

Peeps have special needs, including a brooder with solid flooring, a heat lamp, feeders, waterers, and a ventilated lid.

- **Heat lamp.** Chicks need to be kept toasty for the first few weeks of life, so you'll need to install an adjustable heating lamp over the pen. You can find a range of options commercially, or you can create your own using an incandescent light bulb, which provides some heat.
- **Thermometer.** This allows you to keep an eye on the temperature inside the brooder.
- **Feeders and waterers.** Your peeps will obviously need to eat and drink, so be sure to include chick-size feeding and watering dishes. Clean and check them regularly—peeps can be messy eaters.

One tip I can offer from my experience raising peeps: try raising the little ones in a dividable brooder that starts out small for heat-generating huddling purposes but can be expanded as the birds grow. We did this when raising our second dozen birds. We built a 9-by-4-foot pen and divided it into thirds with removable walls, which allowed us to open up sections as the chicks grew. Worked like a charm.

Isolation Pen

An isolation pen is a chicken coop for one (or one with some chicks). It's for birds that need to be separated from the flock for whatever reason—sick, henpecked, grouchy, or just need space. The pen is intended to be a temporary place to stay, so it's not very large. However, it still needs to house all the items an adult chicken needs, including perches, a nesting box, food and water station, and a place to scratch and dust off. (Flip back to chapter 1 for a refresher.)

The idea is to give the hen a quiet place to recuperate but also a space where you can tend to her needs. A small confined-housing-style chicken coop makes an ideal isolation pen for one bird. You could also get away with using an extra-large solid-floored dog kennel, with perches and a small nesting box inside. You never know when you'll need an isolation pen, but it's a good idea to be prepared with one, just in case.

Sometimes chickens need a time-out for their own good or for the good of the flock. An isolation pen can come in very handy.

Planning Your Coop

Now that you know what chickens need in their abodes, and you've learned about the different types of shelters that can accommodate them, it's time to get into the fun part: designing the chicken coop of your dreams! But that dream should start with a well-thought-out plan.

The planning process is important because it helps you make what you've imagined into a reality. Maybe you've always wanted a brightly colored henhouse with flower boxes and an attached aviary. What steps will you need to take to achieve that goal? When you have an idea in your mind, sketch it out, outline the steps, work through the details, and price the parts before you start building—then you'll be well on your way to constructing your coop.

Of course, plans do change. Case in point: When my husband and I built our newest chicken coop, we had to make some significant alterations to our plan. We drafted a beautiful henhouse, complete with a trussed roof and cupola, sloped nesting boxes, manure-removal hatches, and even an attached storage shed for our girls' food and supplies. We devised a budget, priced supplies, purchased what we needed to get the project underway, and started building. And then, building and hardware supplies skyrocketed in price. We realized halfway through that we couldn't afford to construct the roof we wanted, so we improvised and swapped the expensive trussed roof

Do you have a lot of yard space to spare? That doesn't mean you need to use all of it. Start small and grow from there.

- **Convenience.** Plan to put the coop in a place that has easy access. After all, you'll be visiting it daily (if not more frequently), so it should be in a spot that's simple for you to get to.
- **Good drainage.** Look at the drainage patterns in your yard when planning a place for your coop. It should be in an area that doesn't pool with water during the rainy season. The mixture of rain and droppings-laden bedding makes for a stinky, slimy mess.
- **Good airflow, but not too much wind.** The coop location should get decent airflow so that it can sweep away warm, moist air and associated odors. Think about what's downwind from the coop too.
- **Sunshine.** Make sure the coop gets some sunshine. The sun makes everyone happier, and it gives your birds much-needed vitamins.
- **Natural sources of shade.** The coop spot should have shady areas where birds can hide from the afternoon sun. Bushes and shrubbery are great natural shade-makers.
- **Access to water and electricity.** How close is the nearest spigot? Do you have an extension cord that reaches across your yard?
- **Open location.** Trees are nice, but they also camouflage the roosting and hiding spots of predators.

Although trees and shrubs provide much-needed shade, they can also be hazardous. Make sure your girls are protected from predators while they graze.

What you'll need to consider above all else, of course, is the *amount* of space you have. That will, in most cases, dictate how many hens you can realistically keep and what kind of coop you can put them in. More space means more options for the chicken coop's type and location. But even if you have only a small area to work with, you'll find a range of suitable designs.

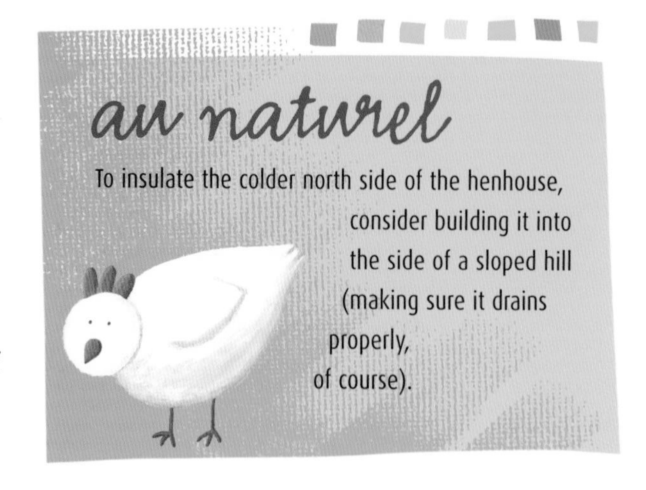

Domestic Resources

A good coop plan includes easy accessibility to all of the things you'll need to take care of your chickens. When the supplies you need are inconveniently located, the whole experience becomes a chore. Position your coop near a garage or storage shed where you can keep things such as hay, pine shavings, the feed bag, cleaning supplies, and the like. In our newest chicken coop, we included a special storage shed just for the hens' goodies. It cut back on chore time and made the whole process more enjoyable.

And speaking of convenience, make note of your coop's water-access options. Are water spigots nearby? Do you need to run pipes out to the henhouse? Do you have a hose that's long enough to reach it? Could you build a rainwater catchment system on the coop's roof? You won't have fun schlepping buckets and barrels of water to a pasture that's 300 yards away, so think about water and how to get it to the henhouse.

If you plan to use heating lamps for warmth or lights for year-round egg production, you should also think about access to electricity. Don't forget to check with your local building department about whether you need a permit.

Screened ventilation ensures that chickens can enjoy a nice breeze without becoming a predator's dinner.

Predator Patrol

Predators play into the planning process in a big way because you'll need to design your coop to keep them out and position the coop where your hens will have the best chance for survival.

As you learned in chapter 1, you'll need to incorporate predator-proofing elements into your coop—secure fencing, locking doors and windows, mesh over ventilation holes, and so on. The extent to which you need to include these features, however, will depend on the critters in your area. For example, if you don't have a problem with raptors in your neighborhood, you may be able to construct an open pen without a wired roof. If diggers are a problem, you'll want to bury 6 to 8 inches of fencing to prevent the varmints from getting into the henhouse.

In addition to including architectural designs that fend off predators, your coop should be located in a place that discourages them. Don't, for instance, put the chicken coop next to a fence that the neighborhood cat walks daily. Instead, put it in a spot away from the fence, where it would be harder for the cat to access. Make it a point to learn more about the predators in your area and incorporate what you need to ensure the safety of your chickens.

The Flock

When it comes to planning your chicken coop, your flock matters, both in terms of its size and its type. Smaller flocks and smaller birds will need less space; larger flocks and larger birds will need more space. But it's not always that easy. Why? Sometimes, more chickens find their way into the family, which means they'll need a large henhouse. Let's look a bit more closely at how your flock will dictate its coop needs.

Chickens are like potato chips—they're addictive and come in countless varieties. Make sure you build or buy a coop that is large enough to grow your flock.

- **Flock size.** As we discussed in chapter 1, birds need a certain amount of habitable space to thrive, so your coop will need to be sized appropriately for the number of birds you have. When planning your coop, you will need to get out a calculator and do the math.
- **Flock type.** The types of birds you have will also affect the amount of space, as well as the kinds of amenities, you need. Itty bitty Bantams, for instance, will need half the space of larger Leghorns. Active birds that like to fly will need facilities equipped for them—meaning fully enclosed, escape-proof pens. Broilers don't need perches while layers do. When planning your coop, examine the specific needs of the birds in your flock and make sure you're addressing them.
- **Flock growth.** Your sweet little chicks won't stay that way, but they're addictive. To be on the safe side, always account for the growth of your flock. If you have the space and the expertise, it's better to build something larger than you think you'll need. Believe me—you'll grow into it. Because layers' egg production slows as they age, you might find yourself buying new peeps each year to keep up with your (or your egg customers') demands. Hens will eventually pass on or—sorry to say—get picked off

Bird Brain

Remember these facts and figures from chapter 1? Well, here they are again to help you calculate the amount of space you'll need for the chickens you want.

- ***Chicks*: Peeps and young birds aged one day to one week need less than 1 square foot per bird. As they grow, their space needs will grow with them. Because they're young and unable to defend themselves, they should be kept in a brooder or a confined shelter, such as a fully fenced-in coop.**
- ***Pullets and young birds*: Chickens aged one week through around twenty weeks can move into an open pen (or one with an accessible yard) that allows 1 to 3 square feet per bird or a confined shelter that allows 2 to 7 square feet per bird. Heavier breeds, such as Jersey Giants, will require more space than lighter breeds, such as Rhode Island Reds; tiny Bantams will require even less.**
- ***Adult hens and roosters*: By the time the birds are twenty-one weeks old, they'll be fully grown. If they're living in an open pen, they'll need 3 to 4 square feet per bird; in a confined shelter, they'll need 7 to 10 square feet per bird. Again, heavier breeds will need more room than lighter ones. An average-size laying hen needs about 6 to 10 inches of linear perch space. One nest box is required for every four to five hens.**

by predators, but your coop's population is still sure to grow.

Something else to consider is where this chicken-centered hobby will take you in the future. As you progress, you may start getting into rare breeds or heritage breeds, keeping them not for their eggs or meat but for pure pleasure. And that means larger or separate coops. Of course, you can't predict the future, but keep all this in mind if hobby farming is something that tickles your fancy.

The Human Factor

Finally, when planning your chicken coop, you have to examine the human factor, which includes your finances and abilities. The type of henhouse and yard you eventually design for your chickens will need to fit within your budget and be easily assembled and maintained by you (and some trusted helpers, of course). Here are some things to think about.

Analyze Your Finances

In today's economy, one of the most important planning details to consider is your budget. Be realistic. How much money do you want to spend on your chicken coop? What can you afford? Before you begin, look at your finances and come up with a working dollar amount (a quick tip: pad that number heavily toaccount for oversights).

These fuzzy little cuties will be full-grown hens in about five months. Make sure to design your coop to have more than enough space because you'll likely get more peeps next year.

Even if the sky's the limit, you'll reach that limit faster than you thought and can find yourself out of money before the project is done.

After you've come up with your chicken coop's blueprints, price the materials before buying them. Head down to your local hardware store or big-box home improvement center and write down how much each piece and part costs, including any tools you need to buy. This may sound tedious, but I highly recommend it. One way to make this chore easier is to chat with one of the hardware store's customer-service folks, who can figure out an estimate for a project in no time.

A coop can include anything you can dream up—even a treehouse topper—as long as you have the budget for it.

Once you've priced your materials, look at your budget again. Have you blown it? If so, make adjustments to your plan. Use lesser-grade plywood, for instance, or swap those shingles for corrugated plastic sheeting. If you've stayed within your budget, great! Proceed with your project. Keep in mind, however, that miscellaneous needs will come up along the way—things you *didn't* plan for. So set aside a little money for incidentals.

When planning and designing your coop, do what you can to make it more affordable. Visit used or reclaimed lumber yards for inexpensive wood and supplies. Scrounge around your own garage and repurpose some materials (for example, my friend used an old bathroom vanity as his henhouse, and a coop-tour participant used an old dog run for her chicken coop). Of course, if you repurpose materials, make sure they're safe for your ladies.

Time and Skills

Another human factor relates to your time and skills. How much time do you have to manage the coop, and what is your ability level? (Be honest!)

First, let's talk time. The larger the flock, the larger the coop. The larger the coop, the more maintenance it will need. If you're like many people today, time is tight. Work commitments, school commitments, sporting matches, dance lessons—with all of those time thieves encroaching on your schedule, you may not have much time left over to tend to your flock. That's okay; just account for it in your coop's design. Here are some features to consider when time is a challenge:

- Construct a smaller coop and keep fewer birds.
- Install droppings trays for quicker bedding cleanup.
- Invest in doors that automatically close when the sun sets.
- Consider timers that turn on the heat lamp or lights.
- Install an automatic watering system.
- Keep the birds on a mesh floor for easier cleanup.

Next, let's talk about skills. If you're keeping chickens, you obviously know how to care for them. But how deep is your knowledge, really? If you're a beginner at poultry keeping, make your setup as simple as possible. Start small, and grow as your skills and experience grow. Similarly, if children are helping keep your flock, the coop should be easy for them to access and work with so that it's a positive experience for them. If you're a chicken-keeping maven, however, plan your coop to suit your more extensive experience. You might be ready to grow your six-hen flock to twenty hens and a rooster and have the time (and maybe some extra helpers) to manage it.

Now for the Fun Part...

Now that you know all about the needs chickens have, the coop designs you can choose from, and the planning considerations, what kind of setup are you thinking about for your girls? Yes, it's a lot to think about, but don't get overwhelmed! Go through the following questions one by one, write down your answers, and use them to come up with a goal-oriented plan that will help you achieve your chicken-coop dream. (Before you forge ahead, though, remember that you first need to determine if you're allowed to keep chickens and have the right climate to do so.)

Make sure you feel comfortable with your DIY skills before you start any project.

- How many birds do you want to keep?
- What types of chickens do you want to keep?
- What predators exist in your area, and how will you protect your flock from them?
- What kind of coop do you want to have?
- What size footprint will your coop occupy (based on the answers above)?
- Where do you want to put your coop?
- What special features will your coop have (for example, insulation)?
- How will you build your coop (for example, will your friends help you)?
- How much money do you have to spend on your coop?

Even though you haven't begun building yet, having a rough sketch of your chicken coop is helpful. It gives you direction, momentum, and something tangible to work with as the project unfolds.

In chapter 4, we go more into depth about the nuts and bolts of your building project. For now, however, it's helpful (and fun!) to jot down a chicken-coop wish list—a rundown of the must-haves and the things you'd like to see in your hens' housing when the project is complete.

First, gather your ideas and do some research. If you're like a lot of chicken keepers, you've kept a curious eye out for coops, looking at premanufactured models at the local feed store, driving through the country to see what other people have built, reading through this and other chicken-keeping books or magazines, or even

SQUAWK BOX

"The ideal henhouse is easily cleaned, provides shelter from the elements, and protects its occupants from predators and rodents. It should be roomy, well insulated, and well ventilated."
Hilary Stern, DVM

This coop not only offers plenty of amenities for its inhabitants but also adds colorful character to the backyard.

going on chicken-coop tours to get inspired. Next, collect everything you've written, clipped, printed, learned, and dreamed about, and make a two-column list. Label one column MUST and the other column WISH. Then, start filling in the columns with your ideas. These could include structural features, such as horizontal perches over droppings boards for easy manure gathering. They could also include aesthetic features, such as bright red paint with white trim and twinkling tea lights and flower boxes as adornments. Your list could even include functional features, such as knobby wheels and comfort-grip handles on a chicken tractor. Just write down everything you can think of. The more detail you have, the easier it will be to construct a budget and see how many "wishes" you can afford.

Finally, draw a rough sketch of your coop based on the items on your list. Include as many details as possible because you'll use them later, when designing and decorating your ladies' home-sweet-home. Of course, plans will change—particularly the details and finishing touches. But the process will get you thinking about the next phase of construction: building the coop.

the To-Do ☑ List

Here's a recap of all of the tasks I've added to your to-do list throughout this chapter.

- ☑ Look at extreme temperatures and write down the architectural features you'll need to include in your chicken-coop design.
- ☑ Check with your local department of agriculture and your homeowners' association or landlord (if applicable) to find out how many chickens you can keep.
- ☑ Invite your neighbors over for a barbecue, serve them grilled teriyaki chicken (seems appropriate), and share your plans for a chicken coop.
- ☑ Survey your property to scope out the best location for your coop.
- ☑ Make note of all your storage, water, and electrical resources and come up with solutions if needed.
- ☑ Call your local extension or wildlife management office to learn more about chicken predators in your area and incorporate tips from chapter 1 into your design.
- ☑ Figure out what types of birds you want and how many you want to keep, and then do the math to calculate the minimum amount of space your flock will require (see page 56).
- ☑ Create a budget for your chicken coop, price the materials, look at your budget again, and make adjustments as needed.
- ☑ Take a realistic look at your time and skills so that you can design a coop that's appropriate for your situation.

The Nuts and Bolts

With your wish list in hand, a reasonable budget sketched out, and a general idea of how you want your chicken coop to look, it's time to dive into your building project. Whether you're constructing a small enclosed pen from the ground up, retrofitting an existing structure, or building a large, top-of-the-line henhouse with a fenced yard, you'll need some plans, some materials, some tools, and some know-how to get started.

Rather than describe inspirational or unique coop designs, this chapter is going to give you the knowledge you need to create your own. First, we'll go through some basics about building materials and tools, such as lumber, hardware, hammers, and saws. Then, we'll discuss repurposing existing structures, including what to look for and how to prepare such a structure for your birds. In the next chapter, you'll find several simple plans that you can easily modify for your property.

Never built anything before? It's not terribly difficult, but you should have another "how-to-build" resource (such as a more detailed book or an experienced builder) on hand for pointers as you journey through your project. And always remember to use common sense when handling tools and heavy materials. This is supposed to be a fun construction project resulting in a home for your ladies—not a trip to the emergency room!

The Building Blocks

As with any project, the right materials and tools make all the difference. They are the building blocks that will determine not only how the finished product will look but also how you get there. A henhouse made with top-quality tools and lumber, for instance, will look and function better and be easier to build than one made with rusty equipment and pieced-together scraps from an old construction job. Both will serve their purpose—housing chickens—but the better quality one will last longer and will result in more efficient use of your time and skills.

In this section, we'll introduce you to some building basics and help you determine which materials and tools will work best for your coop project.

The Materials

Been to a home-improvement store lately? The choices in lumber, hardware, and accessories can be daunting, particularly if you're not construction-savvy. The good news is that you're going to go into the store with a parts list, which is a laundry list of materials you'll need to get started. (Inevitably, you'll have to go back for *something*. It's part of just about any construction project.) The list will include quantities and measurements for specific items, such as lumber, plywood, various types of screws and nails, and the particular brackets and hinges that you will need.

Before you go shopping, you must understand all of the different pieces and parts so that you can choose the most appropriate items for the job. You also must keep your budget in mind when choosing items. Sure, you love your chickens, but they don't necessarily need that striking walnut overlay on their nesting boxes or redwood siding on their entire henhouse. Let's take a closer look at the materials themselves.

If you're not sure you've found the right part, show your list to an employee, who will be happy to help you find what you need.

Always go into a lumberyard with specific measurements and grades in hand. Employees can often make the necessary cuts for you.

Lumber

All those lengths of 2x4s and 1x2s and 2x12s—that's lumber. Lumber is used for structural applications, such as walls, floors, and ceilings, depending on its dimensions, grade, and moisture content. It's also used for fencing, nest boxes, and perches. Lumber is milled from strong softwoods, such as pine, cedar, and redwood, which hold nails well and are easy to cut and manage. Here are some guidelines when choosing lumber for your chicken coop:

- **Use the right grade.** Lumber is graded as *clear*, *structural* (*Select*, *1*, *2*, and *3*, with *Select* being the best quality), *construction*, *stud*, and *utility*. You'll mainly be using structural- and construction-grade lumber.
- **Choose quality pieces.** When selecting particular lengths of lumber, look at them carefully to be sure they are straight and true. Avoid pieces that are warped, cupped, twisted, crooked, split, or checked—their integrity could be compromised, which could affect the strength of your structure.
- **Stay away from pressure-treated lumber.** Some lumber is treated with chemical preservatives to slow wood decay. Because some of these chemicals can be harmful to birds, use untreated lumber instead.
- **Go with exterior lumber.** If your budget allows, consider using naturally insect- and decay-resistant redwood or cedar lumber.

When you're choosing the wood for your chicken-coop project, remember that your decision will affect the durability and attractiveness of the final product. Some woods are more prone to warping than others, some are more resistant to decay than others, and some hold a coat of paint better than others. In addition, matching styles and wood varieties will help create a common theme.

Plywood

Those big flat sheets of wood—that's plywood. Plywood is made up of thinly sliced layers—or *plies*—of wood. The sheets come in thicknesses ranging from $\frac{3}{16}$ of an inch to more than 1 inch, and they're graded *A* through *D* depending on the quality of wood in the outer plies. *Finish* plywood has finish-quality wood

Plywood is generally strong and resistant to cracking, shrinking, and twisting or warping.

veneer on one or both sides; *sheathing* plywood has two rough sides and is used for interior or exterior building use.

You'll use plywood in a number of ways in your chicken coop: as roofing, flooring, and siding as well as to construct nesting boxes, pop holes, doors, ramps for easy henhouse ingress and egress, dust-bath boxes, and A-frame and lean-to shelters. It's perfect for anything that calls for a flat sheet of wood.

When choosing plywood, just as when you're choosing lumber, your main concern will be selecting the right sheet for the job. Finish plywood, for use indoors or in situations where aesthetics matter, comes in grade *A-C* (veneer on one side) or *A-A* (veneer on both sides). Sheathing plywood is grade *C-D* (also called *CDX*) and is rated as either *exposure 1* (for use in areas where some moisture is present) or *exterior* (for use in areas that will be permanently exposed to weather). Sheathing plywood also carries a thickness rating and a roof/floor span rate in inches for roofing and flooring applications. Need help? Talk to your lumberyard and hardware store employees. If you tell them your plan, they'll be able to tell you what type of plywood you'll need.

Hardware

Hardware refers to all of the nails, screws, staples, hinges, and brackets that hold the pieces of lumber and plywood together. The variety of styles and sizes means there's a right piece for each and every job. Here's what you need to know about basic hardware:

- **Nails:** Nails are identified by their typical purpose (flooring, roofing) or by a physical feature (galvanized, coated, spiral). Nail lengths are described by "penny size," or a number from 4 to 60 followed by the letter "d," which stands for the Roman coin *denarius* and was used historically as the abbreviation for "penny" in the United Kingdom. (This wacky system of measuring nail size began in England in the fifteenth century, when "penny size" referred to the price of 100 nails. The larger the nails, the more pennies they cost.) Some popular nails for coop building include *common* and *box* nails, which are used for general framing work, and *finish* and *casing* nails, which are for finish work and fastening doorjambs and exterior trim.
- **Screws:** Nails are preferred for framing and finish, but screws are your go-to fasteners for most everything else. You'll use screws to hang wallboard, install blocks between studs, and attach sheathing and flooring. Screws are categorized according to length, slot style, head shape, and gauge. Popular screws for coops include *wallboard* screws and *deck* screws, both

Buy stainless steel or galvanized hinges, handles, and barrel-bolt latches.

of which drive easily with a drill or screw gun, don't require pilot holes, and seldom pop up as wood dries.

- **Hinges:** If you have doors or lids in your chicken coop, you'll need hinges. A hinge is a joint that fastens two things—such as a pop door to the chicken coop's wall—and allows one of them to pivot. Hinges come in a variety of sizes, styles, and materials for interior and exterior use.
- **Brackets, plates, irons:** These handy pieces of hardware essentially provide extra support to wood joints. They come in myriad sizes and shapes—such as T-plates, angle brackets, and corners—for whatever your project entails.
- **Locking devices:** You'll need to lock your ladies inside at night, so hooks, latches, and locks will be part of your coop design. When selecting these parts, remember that some predators—such as raccoons—can figure out all sorts of locking mechanisms without too much trouble.

Because your chicken coop will be outside, exposed to the weather, invest in good-quality

Using carabiners on latches makes them that much more difficult for intelligent raccoons to jimmy open.

hardware that's designed for outdoor use. When available, choose galvanized or stainless-steel pieces, which will resist erosion and rust. They're worth the investment.

> **SQUAWK BOX**
>
> ***This may be trite advice, but it's certainly worth mentioning and remembering: "Measure twice, cut once."***
>
> ***Hugh Poland***

Insulation

If you live in a cold- or heat-prone area, you may want to insulate your chicken coop. Insulation sits between the exterior and interior walls of the henhouse, helping to isolate the inside air temperature and keep it regulated. You can use all kinds of materials, including straw, fiberglass, shredded newspaper, and spray foam to insulate your coop. They all keep the ladies cozy and warm (or cool).

Insulation comes in some basic forms:

- **Blanket.** This type of insulation comes in flat sheets of fiberglass, mineral wool, and plastic and natural fibers held together with paper; it usually comes in rolls or batts. It's a good choice when you're dealing with space that's relatively free from architectural obstructions, such as window frames or doorways that require cut-to-fit insulation.
- **Foam board or rigid foam.** This kind of insulation comes in poly-based boards that can be fixed to unfinished walls and covered with plywood. It provides thick insulating value in a thin package.
- **Loose fill.** In this case, cellulose, fiberglass, and other loose fillers are blown or poured into place. Loose fill is good for adding insulation to finished areas and around irregularly shaped spaces.
- **Spray foam.** This product is sprayed in and, like loose fill, is good for adding insulation to finished areas and around irregularly shaped spaces.

Insulation comes in different forms. Talk to an employee to discover which type is right for your project.

Insulation is rated with an *R-value*, which indicates its resistance to heat flow. The higher the R-value, the better job it does. The amount and type of insulation you'll need for your chicken coop will depend on your climate and your project.

Other Extras

If you plan to add extras to your chicken coop, such as plumbing or electricity, or any other special features, head to the library for expert information or chat with someone skilled at those specific crafts (or a specialist at a home-improvement store) for details about the right parts and materials for your particular project. If you're not confident doing something yourself, you can always hire a professional.

Warm air rises, and it holds more moisture than cold air. In fact, for every increase in air temperature of 18 degrees Fahrenheit, its water-holding capacity doubles.

The Tools

The tools you use to build your hen pen are just as important as the materials that go into it. Basic carpentry tools, which are the ones you'll use for constructing your coop, are designed to make the building process easier on you so that you can get the job done more efficiently. Plus, they're fun to play with (safely, of course).

You likely have some, if not all, of the tools you'll need in your shed or garage. If not, invest in the best quality tools you can afford. In most cases, better quality tools will last longer and perform better. Before using your tools—old or new—make sure they're in good working order and you know how to operate them safely.

Listed below the bare essentials you'll need for building your chicken coop and a little bit about each item:

- **Chalk line.** When you need to mark a straight line on a flat surface or to mark sheet goods and lumber for cutting, the chalk line's your tool. It's easier and more accurate than a pencil. You just pull the line from the chalk-filled case, stretch it taut, and pluck (*snap*) it using your thumb and forefinger.
- **Circular saw.** A favorite tool of build-it-yourselfers, a circular saw gives you the ability to cut through all kinds of materials quickly and efficiently. The 7¼-inch- and 6½-inch-blade standard-drive circular saws are two of the more popular choices for at-home carpentry use. When it comes to blades, invest in a carbide-tipped combination blade and a panel blade for cutting plywood. Always use caution—meaning proper operation, protective goggles, gloves, and other safety measures—when wielding power tools such as circular saws. And if you're not sure how to safely use one, seek a professional for advice.
- **Drill with basic bits.** Hurrah for the power drill! This tool is one of the most versatile in your tool chest. You'll use your drill to carve holes; drive and remove screws, nuts, and bolts; and even power an electric sander or paint mixer. An all-purpose drill that's standard for DIY projects such as your henhouse is the ⅜-inch (which refers to the

diameter of the bits and accessories that the drill will accept). There are many types available; choose the one that's right for you and your skill level.

POWER DRILL

- **Extension cord.** Because many power tools require electricity, you'll need a power source, so another must-have item is an extension cord. Extension cords come in various lengths, thicknesses, and functions. When selecting a cord, choose one that can handle the power you require and that's rated for outdoor use in wet areas.
- **Hammer.** You'll use your hammer for driving nails and pulling them out. When it comes to selection, it's all about comfort. A hammer should feel good in your hand but be heavy enough to do the job. Hammers come in a range of styles. Your best bet for coop building is a basic 14- to 18-inch framing hammer with a 20-ounce head or a smaller finish hammer with a 16-ounce head and your choice of handle.
- **Handsaw.** Though you'll likely use your circular saw for most of the lumber and sheeting cuts, it's a good idea to have a handsaw handy. There are many types of handsaws for all kinds of woodworking projects, but a good all-purpose carpentry saw is a crosscut saw with 8 to 10 wood-slicing teeth per inch.
- **Level.** If you're building a chicken coop, you'll absolutely need a level. It helps you build walls and coop features that are perfectly vertical and horizontal. Most levels have one or more bubble gauges (horizontal, plumb, and 45-degree angles) that indicate the level's orientation in space at any moment. As the level tilts, the bubble shifts accordingly. For coop building, you'll need a 2-foot carpenter's level as well as a small tool-belt-size version.

SMALL LEVEL

- **Sander/sandpaper.** Sanding smooths and shapes wood in preparation for painting and finishing. In a chicken coop, it's particularly important to sand down rough areas and edges that could splinter or otherwise harm chickens' feet. Belt sanders are used for larger areas; finish sanders or sand-by-hand pieces of sandpaper are used for smaller or more intricate jobs. Sandpaper comes in different *grits*, or levels of coarseness; use the coarser grit for serious sanding and the finer grit for finishing touches. The same safety spiel as mentioned for the circular saw holds true for a sander: use it correctly, wear protective gear such as a dust mask and eye protection, and consult an expert for advice if needed.
- **Screwdriver.** Screwdrivers, whether manual or power, are must-haves for carpentry projects. In general, screwdrivers have either a slotted or Phillips head. Plan to have at least one of each. When purchasing screwdrivers, look for those with hardened-steel blades and easy-to-grip handles. To save time (and your wrist), consider a cordless power screwdriver with slotted and Phillips bits.
- **Square.** A 90-degree square will be an indispensible tool when building your coop. Simply put, it helps you mark your lumber and plywood for cutting, and it also serves as a straight edge when razoring through paper or sheet goods. You'll discover various types of squares; for simple carpentry, a framing square works well.
- **Staple gun.** Like a souped-up Swingline, staple guns power heavy-duty staples into

POWER SANDER

lumber and plywood. A staple gun comes in handy when fastening your mesh fencing material to anchor posts or enclosing your chicken coop.

- **Tape measure.** This essential tool of the trade helps you accurately measure the materials you're working with. A good all-purpose version is a 25-foot steel tape with a ¾-inch blade; for larger projects, consider a 50-foot reel-type tape measure.

STAPLE GUN

Of course, your particular job may entail different tools in addition to these. But in general, the aforementioned should get you well on your way to building your chicken coop. To make transporting your tools easier, load whatever will fit into a bucket (you can also purchase an inexpensive multi-pocketed sleeve for your bucket). And remember to take care of your tools, cleaning them after use—they'll last longer and work better if you do.

Fencing

Fences come in all shapes and sizes. They're tall, they're short, they're solid, and they're stretched. Some are plastic, some are wood, some are electrified, and some have barbed wire. No matter what they look like or are made from, they serve two main purposes: keeping things in and keeping things out. If your chicken coop has a yard or run for your ladies, you'll need to build a fence for their safety and protection.

Fencing provides another opportunity to customize your coop to your needs and tastes—and of course to keep your hens safe.

Raccoons are crafty and will try to get into your coop from all angles. Be proactive and thwart their efforts.

Fences around your henhouse and coop should

- be predator-proof—meaning high enough to prevent jumpers, buried into the ground deeply enough to prevent diggers, and strong enough to withstand handy critters such as raccoons;
- have tightly woven fencing material so that chickens can't escape;
- be strong enough to withstand the weather and normal wear and tear (as well as rough-and-tumble hens); and
- be affordable to install and maintain.

Keeping in mind that the following information is a broad overview of fencing, let's look a bit more closely at the different components of a fence and tips for installing one around your chicken yard.

Materials

When designing your chicken coop, you can choose from many different types of fencing material. The kind of fence you build and how you build it will depend on your particular situation. Generally speaking, however, the following are some basic materials you'll need when constructing a safe fence for your hens.

Posts

Fence posts are those vertical parts of a fence that are anchored into the ground and support the horizontal fencing material. When building your chicken coop, you'll need two types of posts: anchor posts and line posts.

Anchor posts are stout posts around your coop's corners and gates that literally anchor the fence down and give it strength and stability. These are typically thick chunks of lumber that you sink into the ground and hold in place with concrete. Because wood deteriorates rapidly when it's buried in soil, wooden fence posts are typically treated with preservatives. Avoid posts coated with oil-based or creosote-based preservatives because the materials and chemicals used can harm your birds' health.

Chickens will entertain themselves with anything in their yard, including unused fence posts.

Line posts are the evenly spaced posts that circle the yard. In a chicken pen, steel posts called *T-posts* do the job inexpensively and effectively. Shaped like a letter T with an arrow-shaped anchor at the base that you drive into the ground, the T-post's main purpose is to secure the fencing material. These types of posts come in various lengths—from 5 feet to more than 8 feet—to suit your particular needs.

Fencing Material

The *fencing material* is the stuff that runs horizontally from post to post. The most common material used for chicken and poultry fencing is woven wire, which is metal wire that's knotted or welded together, leaving evenly spaced openings. Plastic woven mesh is another option, but it's much weaker and should be used only for temporary chicken-housing situations.

Woven-wire fencing comes in different gauges, or wire diameters; the heavier the wire, the more durable—but also the more expensive. It also comes in different styles, all designed for various housing needs. Poultry and garden fencing are the best choices for chickens. *Poultry fencing* has horizontal openings that are narrow at the bottom and wider at the top. *Garden fencing* has evenly spaced openings that resemble the pattern of a chain-link fence.

Chicken keepers tend to use wire fencing for their coops unless they are fencing their property off from neighboring yards.

Use the best materials you can afford when constructing a fence for your chicken coop.

Lightweight and easy-to-handle poultry netting, which looks like thin wire woven in a ½- to 2-inch honeycomb pattern, is another option, particularly if you plan to create a completely enclosed aviary for your birds. But because some predators (raccoons) can muscle their way through it, you may need to shore up netting with a secondary layer of poultry or garden fencing.

Gate

Your chicken coop will need a gate too. You can fashion your own gate out of lumber and wire mesh, or you can purchase a chain-link-fence gate from your local home-improvement center and add predator-proof mesh to the bottom portion of it.

If you build your own gate, your best bet is to consult a quality fence-building book for detailed guidance. However, keep these things in mind:

- **Location.** Place gates in well-drained areas, and make sure they're in convenient and well-lit spots.
- **Size.** Design gates to be wide enough so that you can fit your equipment through. If you plan to drive your tractor in the chicken pen, you'll need an opening wide enough to accommodate it.

While fencing is a necessity, you can still add your own special touches with decorative latches and handles.

Fasteners and Hardware

Fences and gates have their own set of fasteners and hardware; what you use depends on the type of fence you put up. In most cases, you'll use nails, screws, and staples to secure the joints between pieces of lumber and hold mesh fencing to the anchor posts. You'll also need sliding bolts, hooks and eyes, and other locking mechanisms to secure openings. And don't forget the hinges! Choose pieces and parts that are galvanized or stainless steel, which will better weather the elements.

Fencing Them In

Putting up a fence for your chickens doesn't have to be complicated, but it does take some savvy and skill. For expert building tips, I advise a trip to the local library or home-improvement center to find a book specifically on building fences. It should give you explicit step-by-step instructions for putting up a fence fit for fowl friends. Flip to the Resources section for some recommendations.

Using Renovated Materials

You can transform just about anything into a chicken coop. With a little elbow grease and some basic materials, you can repurpose sound, safe structures for your ladies—but you'll need to carefully inspect any future home before letting your ladies loose. Keep these points in mind when looking at recycled housing:

- **Building-material check.** Make sure the structure is constructed of something that's suitable for chickens. Particleboard, for instance, contains glue and may be harmful if the birds ingest it. Creosote-treated lumber can be harmful to a chicken's respiratory system.
- **Hidden-dangers check.** Look for nail points sticking out, sharp splinters, holes where hens could get their legs or heads caught, unsafe paint that could be pecked, and any other potential dangers.
- **Soundness check.** The structure should be structurally sound, meaning it should be well built and not in dilapidated condition.
- **Predator-safety check.** Examine the structure to be sure there are no holes or other easy-entry paths for creeping or crawling predators. Repair any that you find.

If the structure seems usable, go for it! Build some nesting boxes and install them. Construct a mobile perch and put it inside. Assemble an enclosed pen and attach it to the structure. Some of the cleverest coops are the ones that have been made from repurposed materials. Have fun with it!

Let's Get Building

Now that we've covered the nuts and bolts (literally), you have all of the knowledge you need to start building your coop. The next chapter outlines some basic plans for a confined coop, a mobile coop, a brooder box, a nesting box, and a portable perch. These plans are just the starting point for you. Take everything you've learned so far and transform the basic concepts into the coop of your dreams.

Now that you've learned all about coops, it's time to start constructing your own!

Building Your Coop

As the chicken-keeping hobby flourishes, creative hen-house plans—free or for a fee—are becoming easier and easier to find. An online search will yield scores of pages that have all sorts of super-cool coop blueprints.

Not to be outdone by those rich Internet resources, I thought I'd share my favorite basic how-to plans and drawings. The difference: these plans for hen-houses, furniture pieces, and hides have been tested and approved by our feathered ladies, and my husband and I have had no complaints from them so far!

If you need a little jumping-off point for your poultry-house project, this chapter is for you. It contains detailed

Important!

Keep the following tips in mind for all projects:

- Always predrill holes for screws using a ⅛-inch drill bit to prevent the wood from splitting.
- Make sure to place your screws so that they do not interfere with each other.
- Before cutting, always measure and mark your cuts using a tape measure and pencil to prevent costly mistakes.
- I cannot say this enough: be very careful when using saws and power tools!
- Once you're finished with a project, inspect your work to make sure no screws or staples are poking out. If you end up with rough edges, sand them down by hand.
- Make every project your own, looking back at your wish list for inspiration.

step-by-step instructions with parts lists that you can use when developing your own unique chicken coop. The plans are simple idea-prompters that are easy to personalize with items on your wish list. Now, let's get building!

Building Furniture

As we discussed in chapter 1, furniture in your chicken coop includes pieces such as nesting boxes and perches. With a few tools, some materials, and some time, you can build coop furniture yourself.

Nesting Boxes

Your hens will need a cozy space to lay their eggs. Here's a design for a double nesting box that's very simple to build. You can expand this to as many nesting boxes as you need, as well as include hatches at the top or back that make egg collecting easy. Remember that each box should measure approximately 12 inches wide by 14 inches high by 12 inches deep for an average layer.

TOOLS AND MATERIALS

- One sheet of ¾-inch construction-grade plywood, 48 inches by 96 inches
- One 22½-inch length of 1x2 pine
- Two 12-inch lengths of 1x2 pine
- One 24-inch length of 2x2 pine
- Approximately fifty 1½-inch galvanized deck screws
- Electric screwdriver
- Tape measure
- Pencil
- Circular saw or handsaw
- Electric drill

You may think that lumber dimensions are pretty straightforward, but you'd be surprised. A 2x4 is 2 inches by 4 inches when it's cut, but the drying process reduces it to 1½ inches by 3½ inches. The same goes for plywood: it's slightly smaller than its nominal measurements indicate. So while some of my measurements might seem off, they're actually taking this into account. Keep this in mind as you measure:

- **1x2 = ¾ inch by 1½ inches**
- **1x4 = ¾ inch by 3½ inches**
- **2x4 = 1½ inches by 3½ inches**

How to Build Nesting Boxes

step 1: cut the base and top

For the base and top of your nesting box, measure and mark on the plywood two pieces that are each 24 inches by 12 inches. Carefully use your circular saw or handsaw to make the cuts.

step 2: cut the sides and divider

For the sides and center divider of your box, measure, mark, and cut two pieces of plywood that are each 14 inches by 12 inches (sides) and one piece that is 12½ inches by 11¼ inches (center divider). Set the center divider aside for now.

step 3: attach the sides

Using your scewdriver, attach a side to each end of the box's base, placing one screw at each corner at least 1 inch from the edge and one screw in the middle of each side.

step 3: attach the top

Use your screwdriver to attach the top to the box's sides, using three screws for each side.

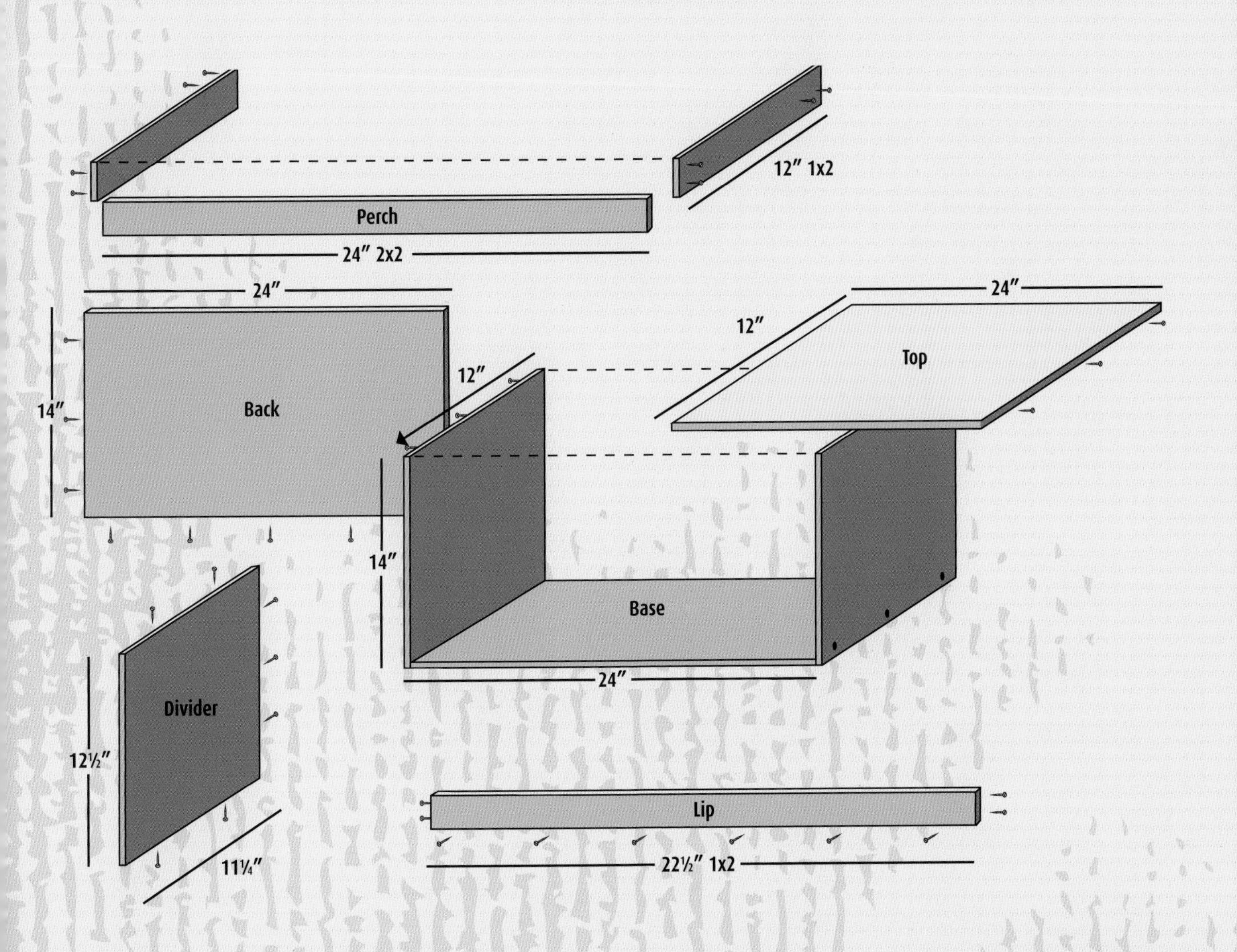

step 4: cut and attach the back

For the back of the box, measure, mark, and cut a piece of plywood that is 24 inches by 14 inches. Secure the piece into place, driving one screw about 2 inches from each corner and one about every 5 inches in between.

step 5: attach the center divider

Lay the nest box on its back and slide the pre-cut center divider piece directly into the center of the box so that it fits snugly against all sides. You should now have two boxes of equal size. Secure the divider into place, driving two screws each into the top and bottom and three screws into the back.

step 6: attach the lip

With the box still on its back, add a lip to the front of the box by attaching a 22½-inch length of 1x2 pine across the bottom front of the structure. Drive two screws through each of the box's sides and into the lip and one through the lip and into the base about every 6 inches in between, making sure the screws drive securely into the plywood and don't poke out at all.

step 7: make the perch

Add a 12-inch length of 1x2 pine to the bottom of each side of the box, allowing 8 inches to extend out in front of the box. Drive two screws through each of these arms into the sides of the box. Connect the two arms with a 24-inch length of 2x2 pine, securing it into place by driving two screws through the end of each arm into the ends of the 2x2.

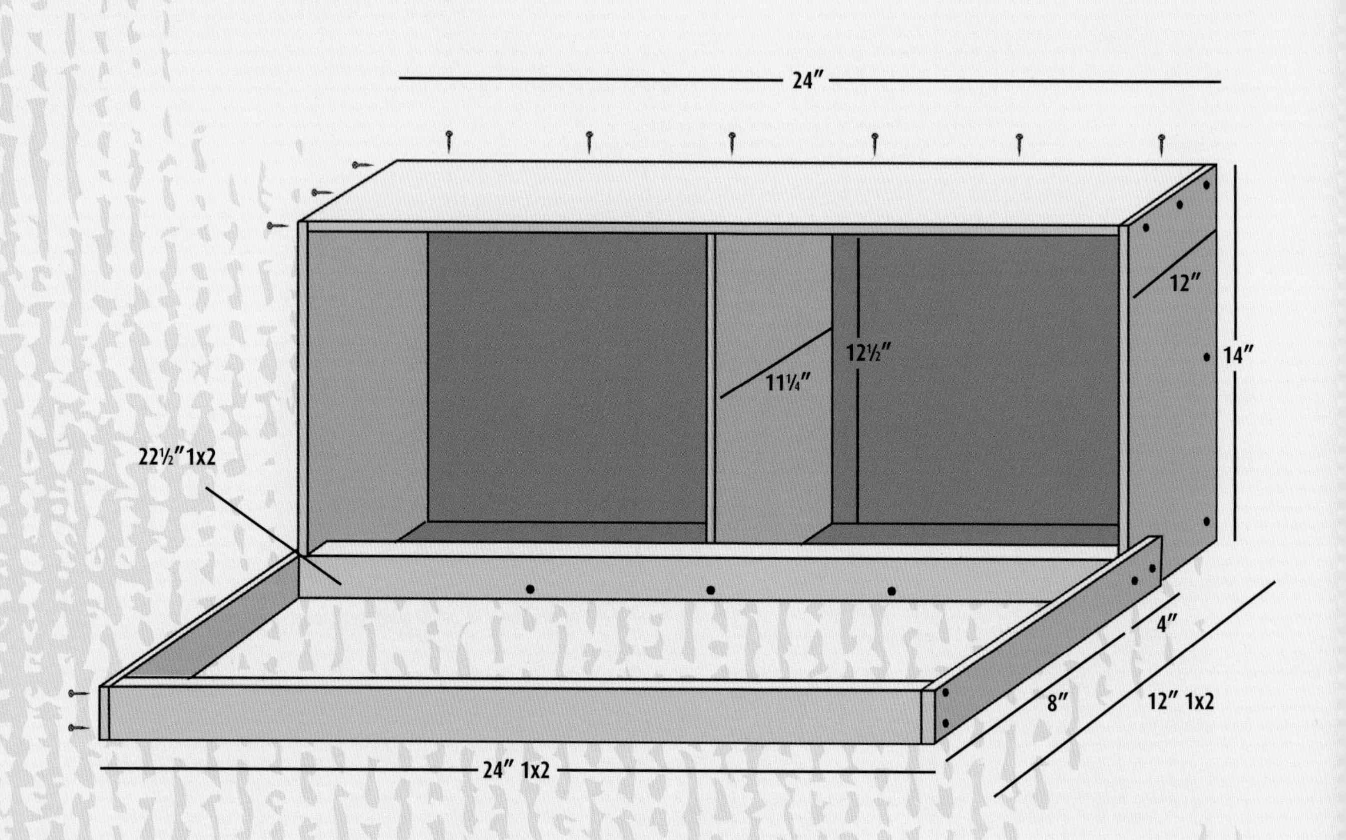

Portable Perches

Your ladies will want a place to roost at night—and here's an easy solution. These simple-to-construct portable perches can be expanded or configured to meet your particular chicken coop's dimensions. Our triangle design is stair-stepped to save valuable space. You'll also find instructions for adding the all-important droppings tray.

TOOLS AND MATERIALS

- Two 120-inch lengths of 2x4 pine
- Two 24-inch lengths of 2-inch dowel
- One 96-inch length of 1x2 pine
- One sheet of ½-inch construction grade plywood, 20 inches by 20 inches
- Approximately fifty 2½-inch galvanized deck screws
- Four 3-inch galvanized deck screws
- One piece of 1-inch poultry wire or mesh, 48 inches by 60 inches
- One handle
- Tape measure
- Pencil
- Square
- Electric screwdriver
- Circular saw or handsaw
- Staple gun and box of staples

SQUAWK BOX

"Chickens are ground-dwelling birds during the day but like to roost off the ground at night. Provide perches at comfortable heights for your breeds of chicken. Make sure that perches are securely attached, not loose or swinging."
Hilary Stern, DVM

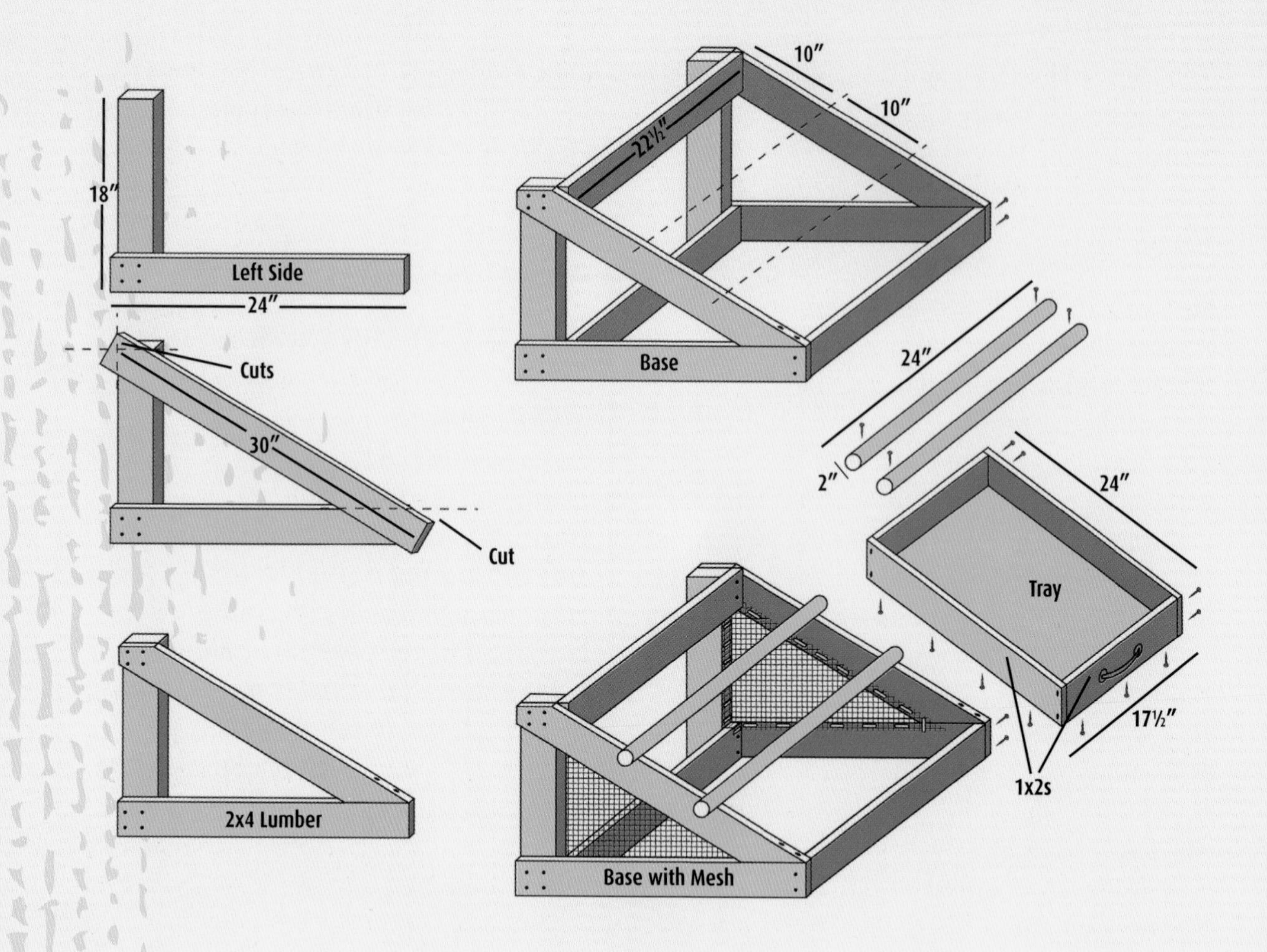

How to Build Portable Perches

step 1: create the sides

Measure, mark, and cut the 2x4 pieces of lumber into two 18-inch pieces, two 24-inch pieces, and two 30-inch pieces. These will make the two right triangles for the sides of your perch. Make a right angle with the 18-inch piece of wood (vertical) and the 24-inch piece (horizontal). Place the 24-inch piece on the outside of the 18-inch piece at the bottom of the 18-inch piece; before fastening, use a square to make sure that you have a 90-degree angle. Attach the two pieces using four screws through the 24-inch piece and into the 18-inch piece. Do the same thing for the other side of the perch, which should be a mirror image of the first side.

step 2: make triangles

Take the 30-inch piece of 2x4 and lay it on top of one of the right angles you've created to make a complete triangle. On the 30-inch piece, mark the angle where it intersects with the 24-inch piece, and mark the two angles where it intersects the top of the 18-inch piece. Cut the angles you have marked using a handsaw or circular saw. The bottom angle should be flush with the top of the 24-inch piece at the front of the perch, and the top angles should fit perfectly on the outside of the 18-inch piece at the back of the perch. Attach the top of the 30-inch piece to the 18-inch piece using four screws through the 30-inch piece; attach the 30-inch piece to the 24-inch piece by sending two screws down through the 30-inch piece and into the 24-inch piece. Repeat this entire step for the other side.

step 3: complete the base

Measure, mark, and cut three 22½-inch lengths of 2x4 lumber; these will attach the two triangles to each other. Screw the 22½-inch lengths to the insides of the triangles at each angle, using two to four screws apiece.

step 4: add mesh

To prevent the ladies from scurrying beneath the perches, block their entryway with wire mesh. Measure and cut two triangles of poultry netting to fit over each side's opening with enough excess to staple. Staple one triangle to the inside of each of the perch's sides, placing a staple on each angle and about every 2 inches in between. Make sure there are no exposed ragged edges that could harm the birds.

step 5: add the perches

On each of the 2x4s that make up the longest sides of the triangles (or the hypotenuse, for you mathletes out there), use your tape measure and pencil to mark two 10-inch increments starting from the top. Make your marks on the top of the frame; these are where your perches will go. You should have about 5 inches left over at the bottom. Predrill holes for the perches, making sure they're level with each other on the opposite sides. Attach the 24-inch dowels by screwing in one side at a time, using one 3-inch screw per side and driving the screw down through the dowel into the frame of the perch.

step 6: build the droppings tray

Build your droppings tray by first cutting two 17½ lengths and two 24-inch lengths of 1x2 lumber. Make a rectangle by attaching the 17½-inch lengths to the insides of the 24-inch lengths, using two screws at each corner (remember to predrill the holes). Measure and cut a piece of plywood to fit on the bottom of the rectangle, and screw the plywood to the 1x2 frame. Attach the handle to one short end of the frame.

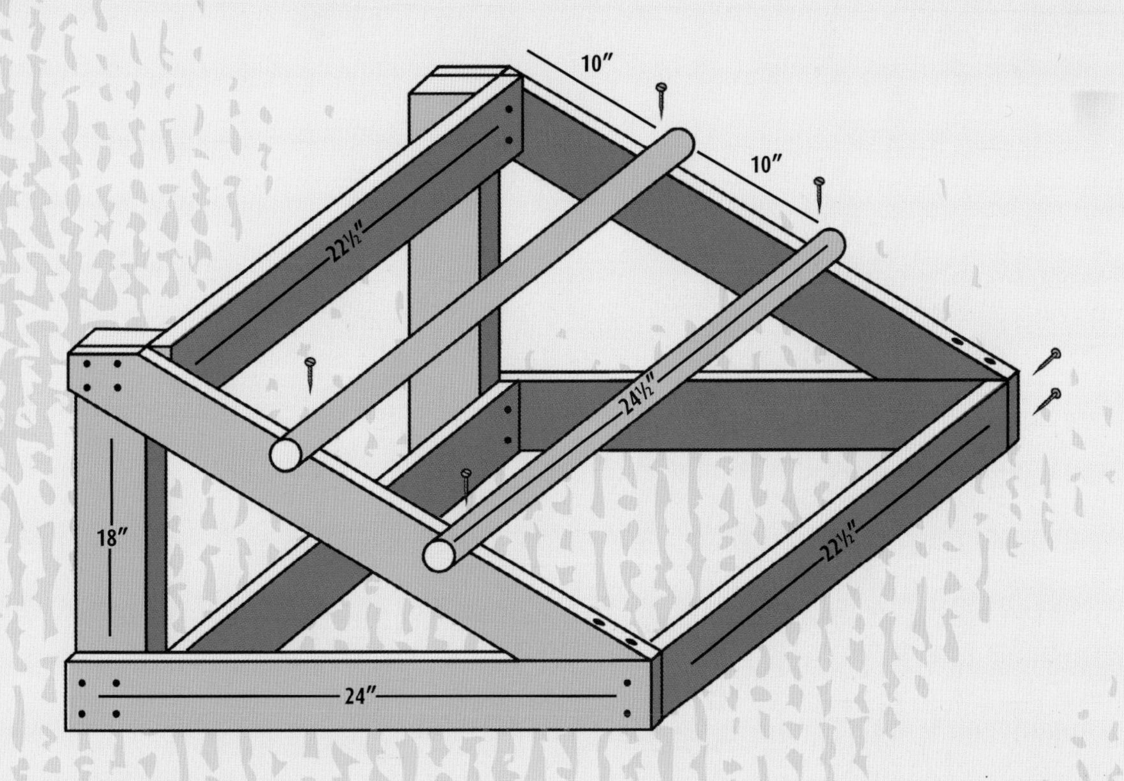

Building Henhouses

Your girls will need somewhere to live, and that's what their henhouse is for. As I mentioned earlier in the book, if you meet certain basic needs, your chickens can live in just about any structure or enclosure. Here, I've included three different henhouses of sorts—an expanding brooder box, a confined pen, and a moveable chicken tractor—that can be adjusted to meet the needs of your flock.

Expanding Brooder

When your chickens are tiny peeps, they don't need much space, but as they grow, their need for room will grow too. If you plan to raise peeps regularly, it's helpful to have a brooder that expands so that you can accommodate them. Here is a super-simple design that's easy to build and easy to clean. Before introducing your chicks, remember to line the box with bedding, install a heat lamp, and provide a feeder and watering station to make them feel right at home.

TOOLS AND MATERIALS

- Two sheets of ⅜-inch construction-grade plywood, 48 inches by 96 inches
- Two 120-inch lengths of 2x4 pine
- One 144-inch length of 1x2 pine
- Approximately 150 ¼-inch galvanized deck screws
- One ¾-inch spade or paddle bit
- Tape measure
- Pencil
- Circular saw or handsaw
- Electric screwdriver

OPTIONAL

- Two 4-inch hinges
- One hook-and-eye safety latch

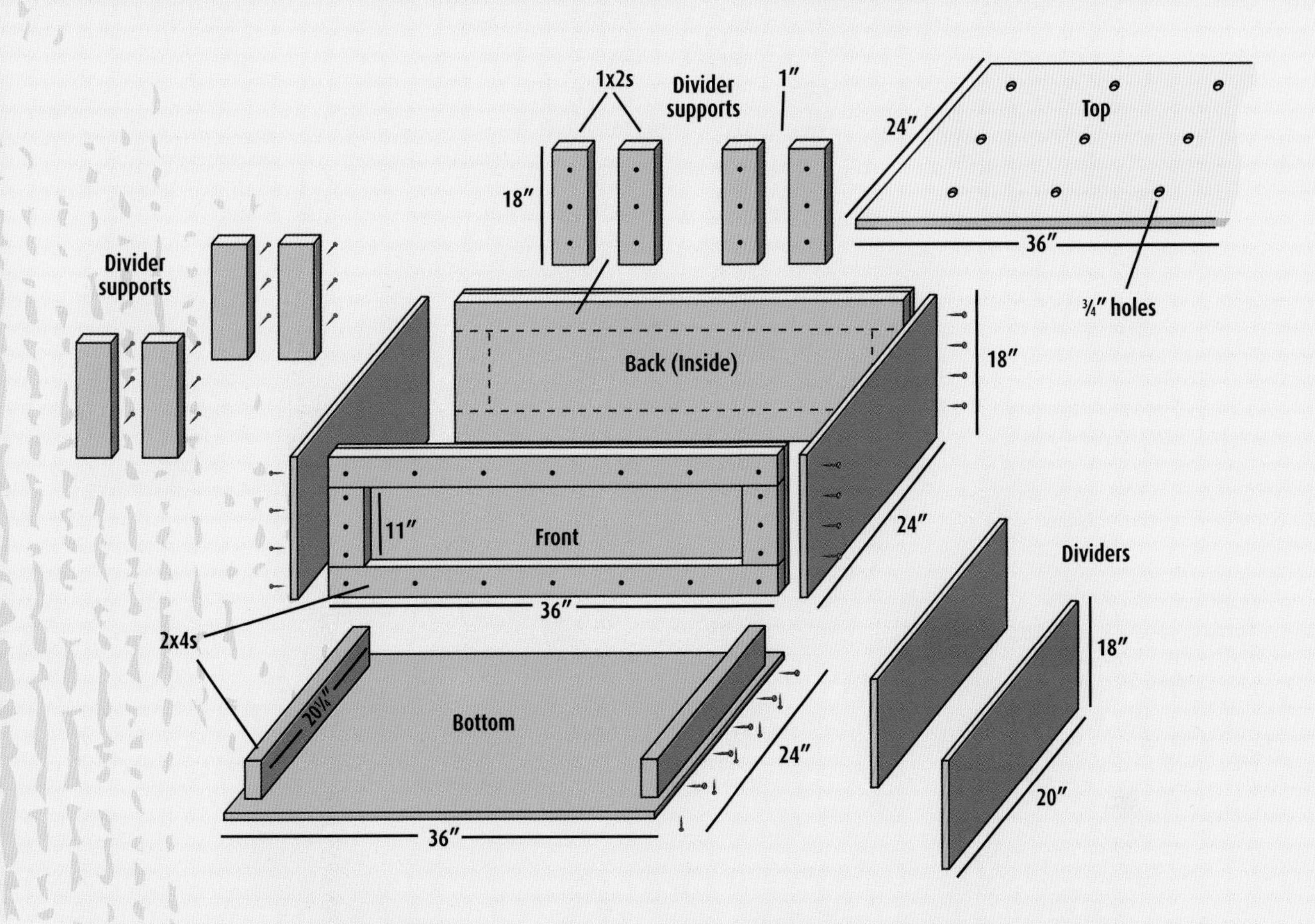

How to Build an Expanding Brooder

step 1: create the pieces

Use your tape measure and pencil to measure and mark on the plywood two pieces that are 24 inches by 36 inches for the top and bottom, two pieces that are 18 inches by 36 inches for the front and back, two pieces that are 18 inches by 24 inches for the sides, and two pieces that are 18 inches by 20 inches for the dividers. Cut the pieces of plywood using your saw. Next, measure, mark, and cut four pieces of 2x4 that are 36 inches long, four pieces that are 11 inches long, and two pieces that are 20¼ inches long.

step 2: construct frames

For the front of the brooder (one of the 18-inch-by-36-inch pieces of plywood), create a frame by screwing two 36-inch 2x4s and two 11-inch 2x4s to it (screw from the inside of the plywood through the frame). Repeat this for the back.

step 3: create a box

Next, attach the sides (the 18-inch-by-24-inch pieces) using four screws at each side, making sure that the corners meet up perfectly. You should now have a box. Note: the frames should be on the outside of the brooder so that the inside of the box is smooth.

step 4: attach the bottom

Lay your 20¼-inch pieces of 2x4 inside the box so that they are up against the bottom of the side panels (the 18-inch-by-24-inch pieces). This creates a frame against the short sides that is similar to the one you created for the front and back of the box. Using several screws for each side, screw through the plywood

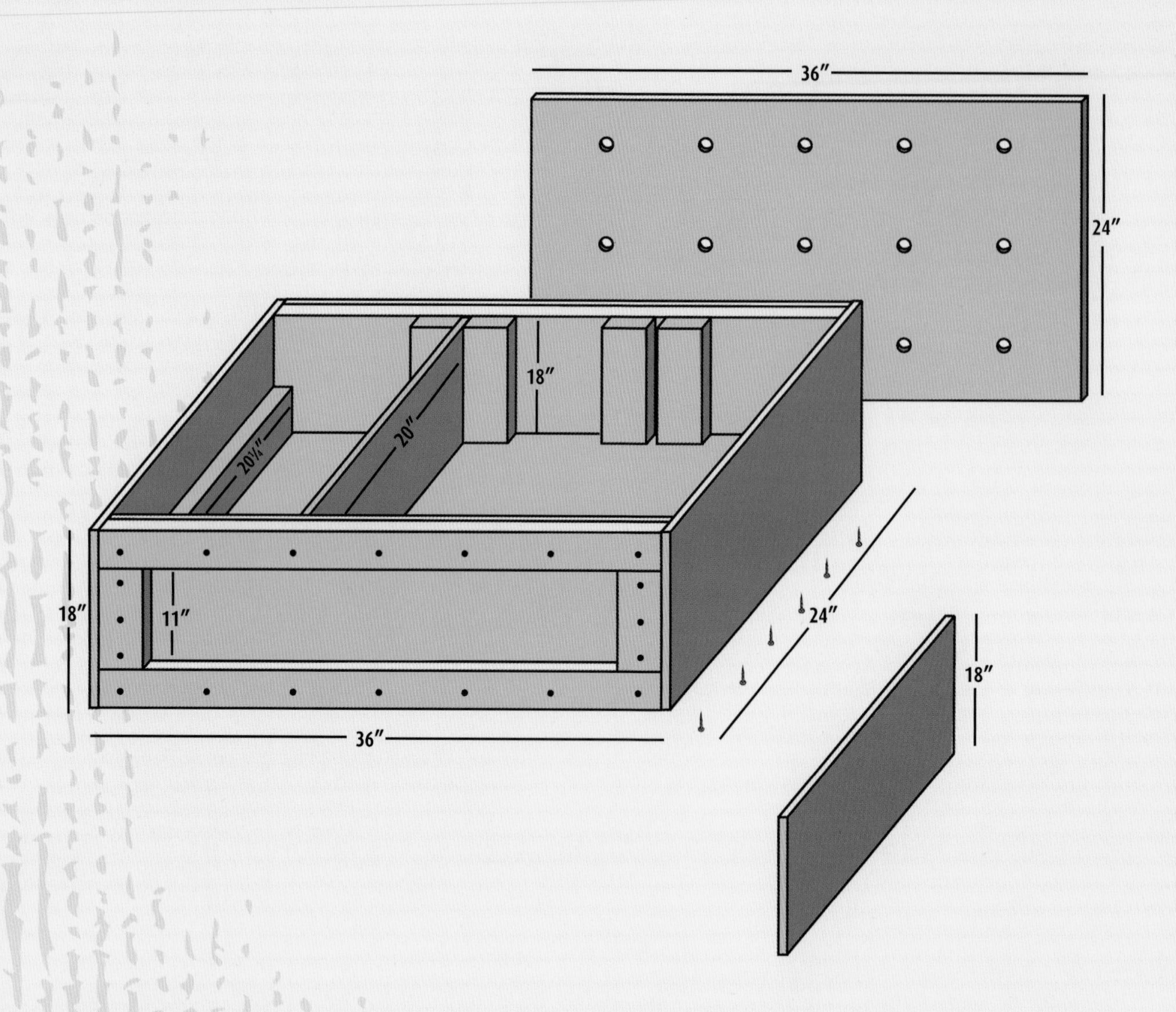

and into the 2x4s. Now turn your box upside down and attach the 24-inch-by-36-inch piece of plywood (the bottom of the brooder) to the frame by sending several screws through the plywood and into the 2x4s around the entire box.

step 5: create the top

The remaining 24-inch-by-36-inch piece of plywood will simply lie on top of the brooder. If you like, you can add hinges and a latch. Drill a dozen or so small holes into the plywood for ventilation using the ¾-inch spade bit.

step 6: create sections and dividers

Measure, mark, and cut eight 18-inch lengths of 1x2 pine. Using your tape measure and pencil, make marks along both 36-inch sides of your box at 12 inches and 24 inches so that the box is evenly divided into three 12-inch sections. At each of the marks, measure and mark ¼ inch out from each side so that you have a ½-inch space between the farthest marks; these marks are where you will attach your divider guides. Attach an 18-inch length of 1x2 vertically on either side of each ½-inch mark (you should have four sets of two 18-inch 1x2s) by screwing the pine into the plywood at the top, bottom, and center. Finally, slide in your plywood dividers.

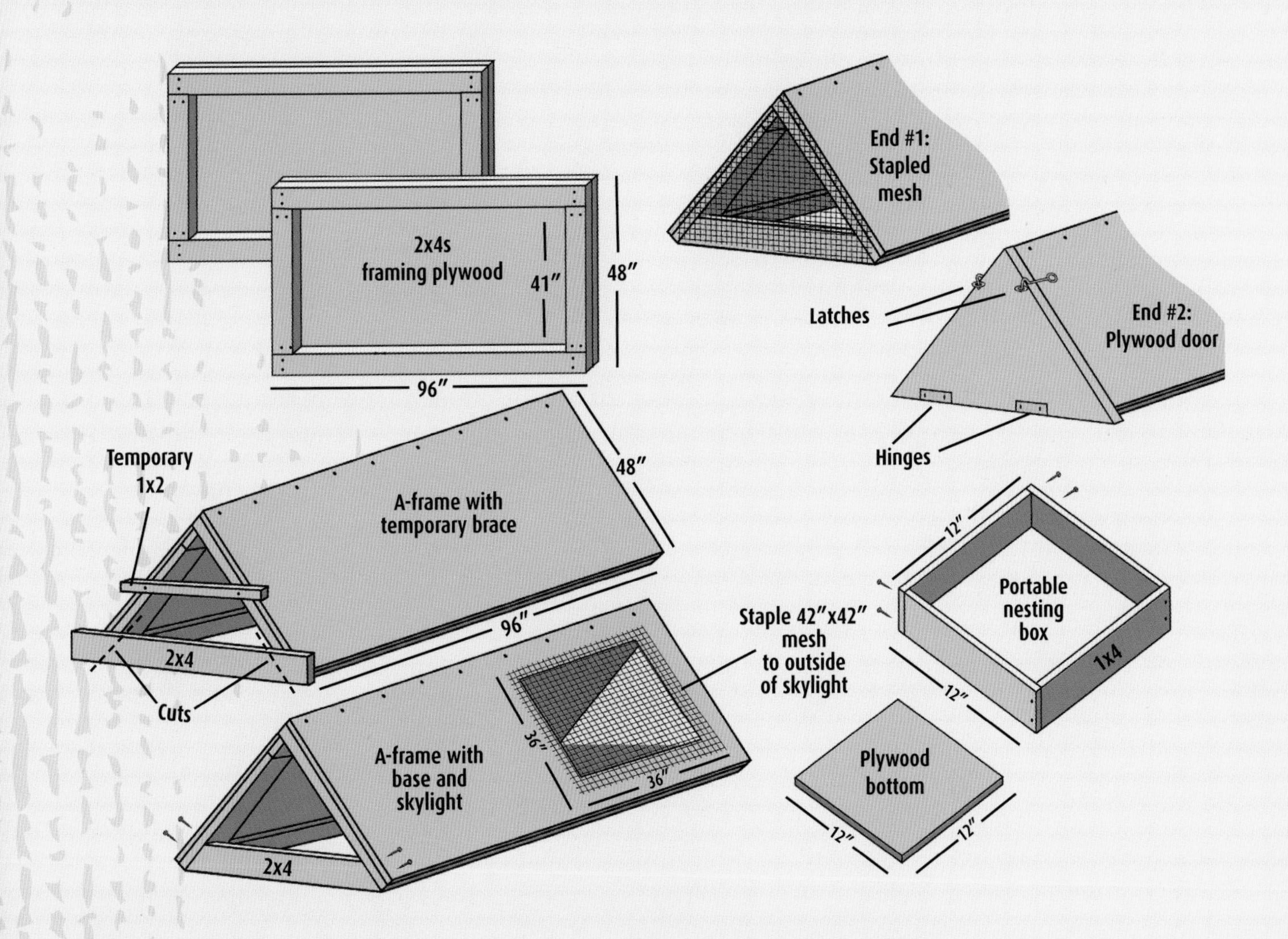

How to Build a Chicken Tractor

step 1: frame the plywood

Similar to the long sides of the brooder project, you're going to first frame your plywood with 2x4s. Measure, mark, and cut two of the 96-inch 2x4s into four pieces that measure 41 inches each. Lay two of the plywood sheets on the ground and screw 2x4s (long side down) around the edges of each sheet, with the 41-inch pieces inside the four remaining 96-inch pieces, placing a screw at each end and then every 6 inches or so. These frames will strengthen the plywood and give you a way to screw the sheets together.

step 2: make the A-frame

Using a helper if needed, carefully lift the two framed plywood boards until they meet at an angle; they should resemble a capital letter A and meet on the inside so that they're flush at the top. Secure the boards together at the A's apex using screws drilled in at both ends and roughly every 6 inches in between.

step 3: add the base

On one side, temporarily brace the boards with a piece of 1x2 that's 48 inches long. Screw the 1x2 horizontally across the opening of the A-frame about halfway between the apex and the ground. Next, place a 2x4 across the bottom of the A-frame's opening and mark the cuts needed for that 2x4 to fit inside the frame. Do this for both sides. Position the 2x4s in place on each end and screw them in so that they're flush with the plywood sheets. Remove the 1x2.

step 4: create skylights

A few inches in from one side of the A-frame, measure and mark off a square that's 36 inches per side. Use your drill to make a series of holes at one corner of the square so that you have a starting point for your handsaw. Using your handsaw, cut out the marked area. Repeat these steps on the other side. Cover the squares by stapling a 42-inch-by-42-inch square of mesh over each opening. Place a staple at each corner and roughly every 2 inches in between, making sure no jagged edges are poking out. These skylights will aid in circulation and give the ladies some sunshine.

step 5: cover one end

Working on the end of the A-frame nearest your skylights, cover the open end with mesh by measuring the triangle, adding 6 inches to each of the three sides for overlap, and stapling the piece to the structure. Place a staple at each point, and then every 2 inches or so in between, trimming away any jagged edges using your wire cutters.

step 6: create a hatch

Now you'll build the access door for the other end of your A-frame. Measure the opening and cut a piece of plywood to fit the space. Attach the hinges to the base of the door and the structure's 2x4 with screws, and then attach the hook-and-eye latches at the top, approximately 6 inches down from the A-frame's apex, one on each side. The door should open from the top, giving you access to the nest box you'll put inside.

step 7: add handles

Placing screws roughly every 6 inches, attach the two 144-inch lengths of 2x4 pine to each long side of the structure's base, leaving 2 feet hanging out on each end for handles.

step 8: make a portable nest box

This removable nest box will simply sit on the ground inside your A-frame. First, cut four 12-inch lengths of 1x4 and screw them together, short side down (this creates the "walls" of your box), into a square. Then cut a piece of plywood to fit the bottom of your nest-box frame and screw it into the 1x4s, placing a screw at each corner and several in between.

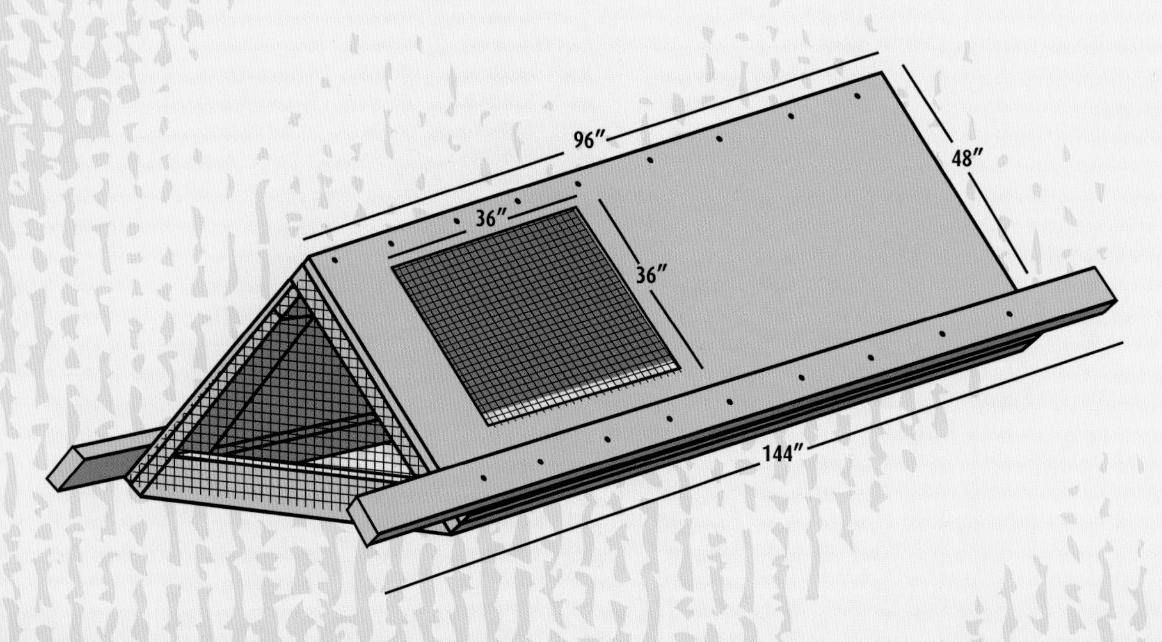

Building Hides

If your chicken coop gives your ladies access to the outside world in a large yard or run, they may appreciate some hides and shaded areas to protect them from flying predators, hot sunshine, or drenching rain. Here are two that work great in our chicken yard.

A-Frame Hide

This simple design gives your chickens shelter whenever they need it. Depending on your yard, this A-frame can be made as large or as small as you need. This particular design is a simplified version of the Chicken Tractor project.

TOOLS AND MATERIALS

- Two sheets of ½-inch construction-grade plywood, 48 inches by 96 inches
- Eight 96-inch lengths of 2x4 pine
- One 48-inch length of 1x2 pine
- Approximately 100 2-inch galvanized deck screws
- Approximately twelve 3-inch galvanized deck screws
- Tape measure
- Pencil
- Electric screwdriver

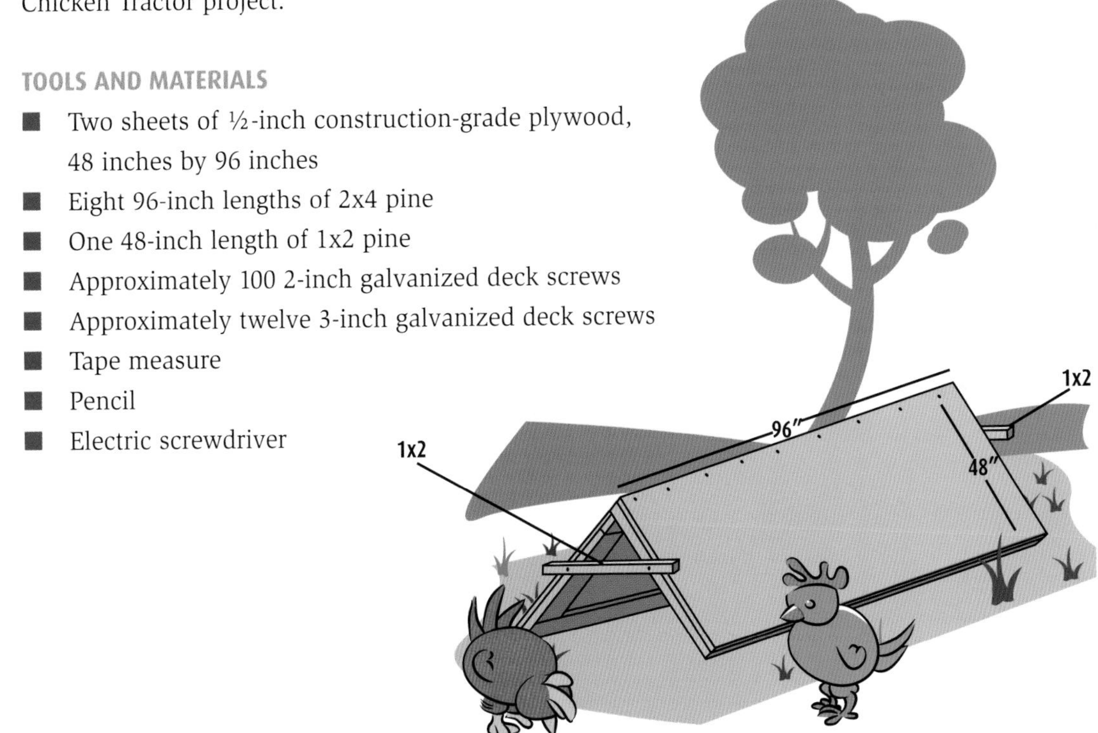

How to Build an A-Frame Hide

step 1: create a frame

Just as with the Chicken Tractor project, you're going to first frame the plywood with 2x4s. Lay the two plywood sheets on the ground and screw 2x4s (long side down) around the edges of each sheet, placing a screw at each end and then every 6 inches or so. These frames will strengthen the plywood and give you a way to screw the sheets together.

step 2: make the A-frame

Using a helper if needed, carefully lift the two framed plywood boards until they meet at an angle; they should resemble a capital letter A and meet on the inside so that they're flush at the top. Secure the boards together at the apex using screws drilled in at both ends and roughly every 6 inches in between.

step 3: add braces

On one side of the A-frame, screw a 48-inch piece of 1x2 into the structure horizontally, about halfway between the A-frame's apex and the ground. Do the same on the other side.

Beyond the Starter Coop

What happens when you want to go above and beyond these basic plans? Build a barn! You can find a wealth of resources at your local library, brick-and-mortar bookstore, or online bookseller that describe everything you need to know about building a larger barn- or shedlike structure. Just remember that those supersized coops require more construction knowledge and skill, as well as building permits and other extras.

Lean-To Hide

Sometimes chickens just need a place to chill out. Lean-to hides work well for that. They provide a place where the ladies can enjoy the warm breezes while being shielded from the sun. This plan is a conversion of the Portable Perches project.

TOOLS AND MATERIALS

- One sheet of ½-inch construction-grade plywood, 20½ inches by 24 inches
- Two 120-inch lengths of 2x4 pine
- Approximately fifty 2½-inch galvanized deck screws
- Tape measure
- Pencil
- Square
- Circular saw or handsaw
- Electric screwdriver

How to Build a Lean-To Hide

step 1: create the sides

Measure, mark, and cut the 2x4s into two 18-inch pieces, two 24-inch pieces, and two 30-inch pieces; these will make the two right triangles for the sides of your perch. Make a right angle with the 18-inch piece of wood (vertical) and the 24-inch piece (horizontal). The 24-inch piece should be outside the 18-inch piece at the bottom of the 18-inch piece. Use a square to make sure that you have a 90-degree angle. Attach the two pieces using four screws through the 24-inch piece and into the 18-inch piece. Do the same thing for the other side of the perch, which should be a mirror image of the first side.

step 2: make a triangle

Lay the 30-inch piece of 2x4 on top of one of the right angles you have created to make a complete triangle. On the 30-inch piece, mark the angle where it intersects with the 24-inch piece, and mark the two angles where it intersects the top of the 18-inch piece. Cut the angles you have marked using a handsaw or circular saw. The bottom angle should be flush with the top of the 24-inch piece at the front of the perch, and the top angles should fit perfectly on the outside of the 18-inch piece at the back of the perch. Attach the top of the 30-inch piece to the 18-inch piece using four screws through the 30-inch piece; attach the 30-inch piece to the 24-inch piece by sending two screws down through the 30-inch piece and into the 24-inch piece. Repeat all of step 2 for the other side.

step 3: complete the base

Measure, mark, and cut three 22½-inch lengths of 2x4 lumber, and then screw them to the insides of the triangles, joining them near each angle using two to four screws apiece.

step 4: add the plywood top

Position the piece of plywood on the front of the hide (where your perches would have gone in the Portable Perches project) and attach it using screws at each corner and about every 6 inches in between.

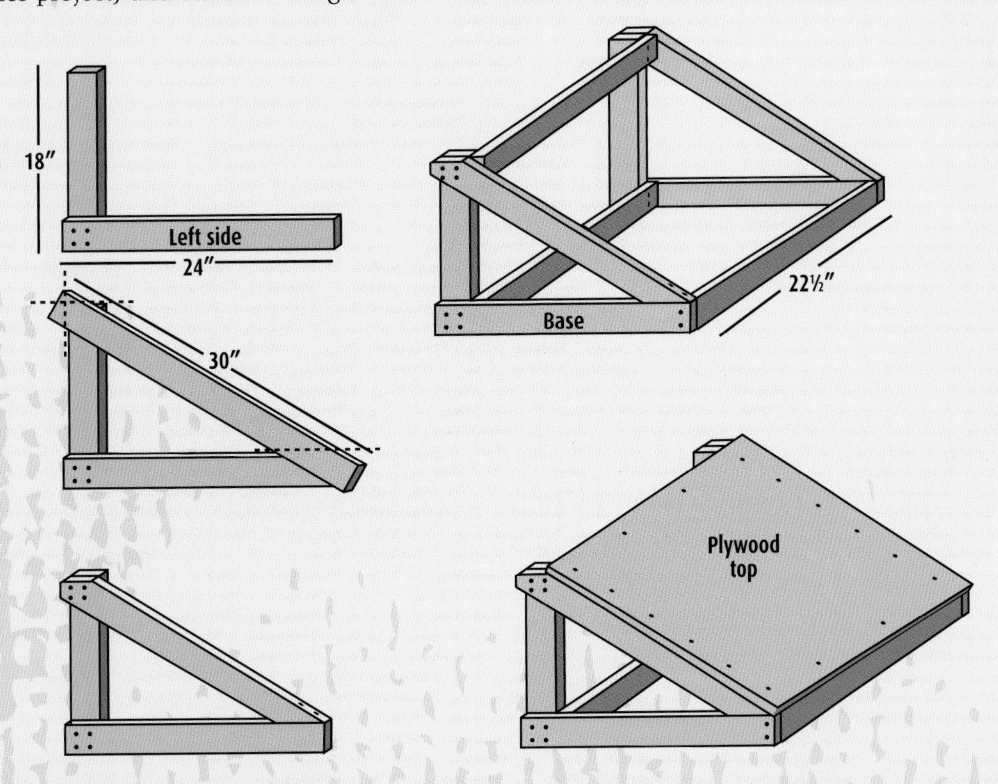

Finishing Touches

Finish work is all about the details. In construction jargon, finish work refers to a building's paint, trim, window treatments, plumbing and lighting fixtures, cabinets, landscaping—all of the aesthetic extras that go above and beyond simply creating the structure. Finish work pulls the project together and transforms the rough work, or the lumber-and-nails part of the building, into a picture-perfect result that's habitable, comfortable, and easy on the eyes. It's the same thing with chicken coops. Those finishing touches make the coop look good in your backyard rather than seeming like a monolithic monstrosity that's taken over your property.

Though your ladies probably won't care about details such as paint color or exterior trim, you should. Finish work not only makes your chicken coop a home-sweet-home for your birds but also improves its durability and can even better the lives of your hens. Paint, for instance, both protects wood and makes the henhouse blend nicely into the landscape. Window coverings insulate the building and the birds against heat and cold while adding that special touch to the structure. These little details are how you put your unique stamp on your coop.

One of my favorite chicken coops was a structure with a cutout facade that was painted to look like a scene from an old cowboy movie. The henhouse's windows—complete with filled flower boxes—were placed along the fake, though convincingly

A wooden bench can double as an exercise and play gym for your chickens. Make sure the wood is untreated.

stenciled, buildings. Against the coop's fencing stood cutouts of a jailhouse, a run-down saloon, and a general store. The door to the store led into the cleverly disguised chicken run, where the chickens had a planted yard and lots of room to stretch their wings. There was even an area for humans to sit under a cottonwood tree and watch the hens play. The owners of that coop created a welcoming environment, and it was all in the details.

So now that you understand what your henhouse and chicken yard *need* to have in terms of their structure and accessories, you can start incorporating all of the finishing touches—such as a hand-painted facade or window boxes filled with petunias—that you *want* the coop to have. Way back in chapter 3, we recommended gathering and keeping pictures and ideas for your dream chicken coop. Well, pull out those ideas, because it's time to get inspired! In this chapter, we'll explore how to include some of those details in your own chicken coop.

For the Birds

Though it's unclear whether birds have preferred pastimes, hobbies, or even places to lounge in the sunshine, they do like to entertain themselves while they go about their day-to-day business of scratching and eating. No one—not even a chicken—likes mundane routines. If your birds get too bored, they may start picking on each other. That's why it's important for the ladies' mental and physical health and well-being to give your hens things that will stimulate and busy them.

If you're keeping your chickens in a chicken tractor or fenced range and you're moving it frequently, the regularly changing landscape should provide the stimulation they need. But if you keep your girls in an enclosed or even yarded pen, think about using these boredom-busting tricks:

- Plant a chicken garden
- Build a chicken play area for climbing and exercise
- Let your birds play with their food

Read on to learn more about how to include these elements in just about any chicken-coop design.

SQUAWK BOX

"One way to provide greens to your chickens is to allow them to 'mow' your lawn for you if your yard is such that you are able to let your chickens wander about safely. Make sure not to use pesticides or herbicides in your garden. And protect your favorite flowers and vegetables from unwanted intrusions!"

Hilary Stern, DVM

Plant a Chicken Garden

When given the chance, chickens will grub on greens and supplement their diets with foraged food that they find in the yard. Though their feed technically contains everything they need to live, the plants provide heavy doses of vitamins and nutrients, which they then share with the lucky humans who eat their eggs. Studies have shown that eggs laid by hens with access to pasture grass or other greens contained less cholesterol and less

Planting a garden specifically for your chickens provides healthy greens for them and keeps them out of your garden.

saturated fat than factory-farmed eggs, and they're loaded with vitamins A, D, and E as well as beta carotene, folic acid, and omega-3 polyunsaturated fatty acids.

If you're eating your hens' eggs, why *wouldn't* you want to pump up the girls' intake of greens? The birds will scratch at the soil and pull up the shoots, and they'll nibble at the established foliage. Hens will also have a heyday with any running water, chasing after it as it drains from the pots. And if you choose to toss the ladies some of your landscaping clippings and kitchen scraps, they'll clamor for those too. It's always entertaining to watch them scramble for the tastiest morsels. (After you've supplemented your hens' diet for a while, take note of the color of their egg yolks. A really cool fringe benefit of feeding your hens fresh greens is that they'll be ingesting high levels of *xanthophylls*—pigments that turn yolks a deep yellow color.)

Here are some plants and green matter that you can toss to your girls or plant in their yards—making sure, of course, that you're using no chemical fertilizers or pesticides.

Ground Covers and Grasses

- Wheatgrass
- Hay
- Legumes (such as alfalfa and clover)

Growing tip: To allow the grass and legume seeds to germinate and grow, divide your chicken yard into sections with a temporary (and chicken-proof) fence, planting one section and leaving the others for the hens. Once the plants have started to grow, let your ladies at them!

Annuals and Flowering Plants

- Kale
- Spinach
- Yellow flowers (such as marigold and nasturtium)
- Sunflowers

Growing tip: A visit to your local garden center will reveal lots of seasonal chicken-safe plant choices. Transplant groupings of different varieties in a low planter box filled with organic soil. Put them in your chickens' run and let them scratch and peck away—you can always put new plants in there later.

Perennials and Shrubs

- Bamboo
- Ferns
- Taller, well-established shrubs such as daphne (the birds will eat the lower leaves)

Growing tip: These types of plants work best in larger yards, where they can be fenced off to grow (and re-grow time and again).

au naturel

Fill your chicken coop's water barrel by harvesting the rainwater from the henhouse roof. Simply run a gutter along the roof and direct the water into the barrel. Alternately, you can direct the rainwater into a barrel kept for watering your chickens' garden.

Build a Play Area

No, you don't need to provide your hens with a huge swing set and slide—though they'd probably have fun with them! A chicken-style gym or play

Shrubs provide shade, shelter, and a great vantage point to survey the yard.

area, built from branches, straw bales, plastic buckets, or even an old dog kennel, gives your girls a place to play, exercise, and hunt for bugs. As long as it doesn't give them an easy escape route, a play gym can add a fun distraction to your chickens' day.

For our chickens, we've fashioned several different play gyms out of thick oak and madrone branches and logs left behind from chopping firewood. We've also used old pieces of untreated lumber, such as 2x4 scraps left behind from—what else?—building chicken coops. The pieces, laid on the ground and strategically overlapped on top of one other, stack like Pick-Up Sticks (remember that game?), and we've made them sturdier (and safer) by screwing them together so they won't topple. The chickens love it. They scratch and peck at the wood, looking for bugs; hop all around it; and chase each other off of it. And the roosters—they love to get on the highest perch and cackle their *cock-a-doodle-doo.*

You'll need to consider some factors when constructing a play gym for your birds:

- **Use chicken-safe materials.** Chickens will peck at anything and everything, so make sure that whatever you use is safe for the birds. For example, don't use pretreated lumber or branches that have been sprayed with chemicals. Stay away from objects with peeling paint. (You get the picture.)
- **Inspect your materials carefully for hazards.** Check them for nails, sharp or jagged points, or other dangers that could injure the birds. And make sure there are no holes for the chickens' feet to slip into.
- **Make sure your structure is sound.** Chickens weigh only a few pounds each, but that weight could cause a poorly positioned play gym to collapse and possibly harm one of them. Your configuration should be solid, sound, and secure.

This barrel provides a nice place to hide from predators or bullying chickens as well as a shady and cool spot out of the sun.

Chickens aren't particular when it comes to decor—a log culled from a fallen tree is enough to keep them happy.

- **Choose an escape-proof location.** If you're going to build a vertical play area, make sure it's not near the fence or where chickens can hop up and fly the coop. Put it in the center of the yard or construct a horizontal area instead.

A play gym or play area gives your ladies a fun place to hang out and do their chicken thing. Plus, it's fun for you to watch their antics.

Have Fun with Food

After spending time with chickens, you'll discover one thing: they love to eat. When it's time to fill the feed troughs, those little ladies come running (some even peck at my muck boots with their tiny beaks!). And inevitably, despite our best attempts to keep the food in their bowl, they turn it over and spill the pellets everywhere—giving them the opportunity to hunt and scratch for their food.

To play on this predictable messy behavior, you can provide different and interesting venues for feeding the birds, making it as entertaining and challenging for them as possible. Make mealtime fun by:

- **Hanging fresh fruit and vegetables.** When designing your food and water station, include an area where you can suspend broccoli or cauliflower heads at the chickens' beak level on a string or a chain with a clip. We use a repurposed suet holder in our hens' yard. I fill it with celery ribs, carrots, lettuce, tomatoes—whatever we're eating—and hang it for the ladies to nibble at all day.
- **Hanging a bird feeder or netting filled with grains.** You could also include a low-to-the-ground bird feeder or netting filled with whole-grain birdseed and give it to the hens as a treat. They'll pick and peck at it (and they'll likely spread the seed all over the ground, just as they do the pellet feed).
- **Feeding grass or hay in a slow-release trough.** You could also stuff hay or grass clippings into a device that makes the ladies pull it out.

Pouring feed into a trough is simple enough, but there are plenty of ways to encourage your ladies to play with their food.

Though your chickens will relish a range of table scraps from your kitchen, you should not let your ladies eat raw potato peels, which are hard for them to digest; strong-tasting foods such as onion and garlic, which give their eggs and meat an unsavory taste; or avocado, as the pit contains a compound called *persin* that can be toxic to chickens (this warning goes for guacamole too). Also steer clear of spoiled or rotten food, fried food, and anything that contains alcohol, caffeine, sugar, sugar substitutes, or high levels of fat—these foods are not good for you, and they're not good for your feathered friends either!

Keep boredom at bay by making your girls work to eat with a hay trough like this one.

Of course, you could simply toss fruits, vegetables, grains, and grass on the ground and let the ladies eat it from there, but they'll be more stimulated if accessing the food is a challenge—and they'll be more fun for you to watch.

For the Humans

Sure, all of those details and extras for the chickens are fantastic for them, but some of the real personalizing and customizing happens when you add finishing touches for you, their human. You've been thinking about this project for some time now, so you probably have an idea of how you'd like the completed product to look.

Some of those special additions may include

- paint, caulking, and other finishes;
- door and window coverings;
- trim around windows, doors, and other such architectural details; and
- fun decor, including signs, twinkly lights, and other whimsical touches.

Looking Good, Coop

Exterior finishes, such as paint and caulking, not only make the coop look good but they also make it more durable and easier to clean and maintain. The paint preserves and weatherproofs the wood while the caulking seals up holes where chilly drafts and precipitation can creep in. These steps aren't essential to your coop design, but they add to its aesthetics and function. Plus, it's fun to choose paint and get creative with color combinations!

But what kind of paint is safe for chickens? Could they get sick from the paint fumes? Your hens' wellness should be of utmost importance. So before buying the cheapest brand of eggshell-white paint, think about some healthier alternatives. You could go with an all-natural water-based paint, such as one of the commercial varieties found online or in your local hardware store. If you have the skills and savvy, you could even mix up a milk- or lime-based paint or wash from scratch. The "green" versions are my first choice, but they can be expensive and hard to find or make. And some can't be cleaned, which is definitely a negative in a chicken coop.

Viable and more accessible alternatives are child-safe low- or zero-VOC (volatile organic chemical) paints and finishes that emit fewer harmful compounds. Studies have shown that exposure to VOCs over time can cause all sorts of health problems for humans, so you can imagine how chickens and their delicate respiratory systems could react! Best bet: go with a child- (and chicken-) safe water-based paint that's easy to clean, such as exterior acrylic.

When it comes to choosing caulking, painting the trim, and adding other such finishes, use materials that are solvent-free, environmentally safe, nontoxic, and nonhazardous. You can find plenty of options that fit these criteria at hardware stores and online.

With a little imagination, some paint, and the right decoration, your coop may just become the talk of the town.

Bird Brain

Lime whitewash is a simple mixture of hydrated lime (calcium hydroxide) and water. It is a tried-and-true method of covering walls—particularly interior walls in barnyard buildings—that has been used for generations. Lime is chicken-safe and inexpensive, discourages the growth of bacteria, repels bugs, brightens up dark henhouses thanks to its light-refractive properties, and is easily tinted with natural pigments.

Doors and Windows

As you read in chapter 6, gates, doors, and windows are part of the main coop construction. You need access to the yard and henhouse via human-size doors, and your hens need access via chicken-size pop holes that most potential predators can't fit through. Screened windows that open and close help with air circulation and temperature regulation; these elements are must-haves. But you can also customize and personalize your doors and windows. Some ideas to consider include:

- **Dutch doors.** These doors, perfect for a larger henhouse, have upper and lower halves that you can open separately. You can open the top to help with air circulation while leaving the bottom closed, protecting the hens from ground-dwelling predators.

This ingenious little hatch allows the ladies out into their run during the day and can be closed to contain them in the henhouse at night.

- **Screened or seasonal doors.** If you live in an area with temperature extremes, you may consider a screened door for summer and a more solid storm-type door in the winter.
- **Automatic pop holes.** If you're not sure whether you'll get home in time to lock in the chickens before sunset, consider automatic doors that open and close on a timer or based on the sunlight. In most cases, your birds will flock inside to their roost before the sun goes down, but keep in mind that there's no guarantee they'll make it in before the doors lock.
- **Windows.** Screened sliding glass windows are a nice option for increasing airflow and bringing sunlight into the henhouse. In the warmer months, you can open the windows and use them as secondary ventilation ports; in the cooler months, south- and west-facing windows welcome warming sunshine—something every creature can appreciate! If your budget allows, consider installing them.
- **Window dressing.** Fun and useful accoutrements to add around your windows (glass or otherwise) include window boxes filled with flowers, shutters to regulate airflow while keeping chickens in and predators out, and awnings to reduce the amount of sunlight that pours in during the warmer months.

Whatever you decide, remember that all doors and windows should be easy to secure.

Get creative with trim colors. It may be your chickens' coop, but why shouldn't you enjoy it too?

All the Trimmings

Trim and exterior facades give the chicken coop and henhouse a finished look. Besides adding interesting detail to the structure, trim frames windows and doors, hiding any holes or imperfections left over from the rough construction process. Trim also gives you a place to fasten your door and window hinges and other hardware.

Though you can use any size or shape lumber for trim (think ornate Victorians adorned with fancy carvings, for instance), the easiest (and cheapest) material to use is a sanded 1x2 slat of wood. You simply cut the lengths to fit and nail them to the building. You can then paint the trim

au naturel

Solar-powered generators attached to lights and other accessories are efficient and low-impact additions to your chicken coop.

to match or complement the structure. Another option is strips of unfinished molding. They're a bit more decorative if you really want to pretty up your coop, but they do come at a price.

When installing trim, have fun with it! Just make sure all nails are well sunk into the wood so that the birds won't harm themselves.

Make It Yours!

The most important finishing touch is whatever you do to make the chicken coop your own. Even if you decide to purchase a premade cookie-cutter variety, you can still personalize it with a themed setup, adornments, and even help from young chicken keepers. The best details come from your own imagination.

A chicken coop doesn't ever have to look like livestock housing. It can be as tidy and cute as you want it to be, and it can reflect your interests outside of chicken keeping. Perhaps you want to make the coop look like a little barn with bright red paint and white trim. Maybe you'd like to construct a chicken-size replica of your own home or design your coop to look like a country cottage with battery-powered tea lights and an English garden. Whatever you choose, create a coop that makes you smile.

Another way to personalize your chicken yard and henhouse is to add adornments, such as clever or personalized signs, outdoor lights, container gardens, or yard ornaments. People who love chickens tend to love chicken decor, so why not include some chicken-shaped planters and signs with sayings such as "What's Clucking?" or "Eggs for Sale"? Give your coop as much curb appeal as your home has.

This sweet little "barn" coop fits neatly into this suburban backyard as an attractive design element.

The style, color, and personality of this henhouse and coop integrate nicely into the backyard.

If you have children, let them in on the fun too. Ask them to help paint the coop. Designate a wall on which they can paint a farm-themed mural or leave colorful handprints. Take them to the garden center and ask them to pick out some plants or yard art to put in the chickens' yard. The more you involve them in constructing the coop, the more interest they'll (hopefully!) show in the hens. And before you know it, you'll have some very handy helpers raising your chickens with you.

From Rough to Finish

The details make a difference in your chicken coop. When the rough building is done, adding those details—paint, trim, a chicken garden, and your favorite metallic rooster sculpture—turns a ho-hum shelter into that chicken Shangri-la you've always dreamed of. It increases your enjoyment in the hobby, for which your hens will show their appreciation. After all, they care more about your attention (and the feed you deliver) than they do the paint color.

Adding extras to your henhouse and chicken coop makes them your own—dear to you and unique showpieces worth sharing with your friends and (hopefully enthusiastic) neighbors. Who knows? Your henhouse could be featured in a chicken-coop tour one day!

Coop Upkeep

Call me crazy, but I love spring cleaning. Though we keep the house clean year-round, there's nothing like scrubbing from eaves to baseboards, knocking down those cobwebs, vacuuming up the dust bunnies, and wiping away all the ash that's floated to flat surfaces from the wood stove. Not only does that deep cleaning give me the opportunity to start with a sparkling slate when the weather gets warm again, but it also allows me to check for items that need repair.

Our chicken coop gets a similar scrub-down each year. We do our daily maintenance chores, such as collecting eggs, freshening up the bedding, and checking for holes in the fencing, but we do the real thorough cleaning and any needed repairs in the spring. That's when we completely change bedding, scrape down droppings, clean and sanitize perches and nest boxes, and seal up any leaks that occurred over the winter. No matter your coop's size or setup, you'll need to do these regular cleaning and maintenance chores too.

Your chicken coop will need both regular upkeep and deep cleaning to prolong its life and give you the most value for your dollar (and labor!). A clean, well-maintained chicken coop will better weather hot sun and damaging precipitation. Droppings won't collect on surfaces and create a smelly mess. The lumber and hardware will last longer and perform better. If you ignore regular maintenance, you'll have a broken-down henhouse in no time.

A clean coop makes everyone happy—the chickens that have to live in it, and the humans who have to look at and smell it.

Additionally, a clean coop is a must for your hens' health and well-being. Birds need clean air to breathe and safe, sanitary areas in which to roost. Dirty, poorly maintained coops can lead to all sorts of illnesses caused by organisms that flourish in warm, moist environments. Chickens could also injure themselves on a protruding nail or broken piece of wire. You like to live in a clean space, and so do your hens.

In this chapter, you'll learn what chores you need to complete, when you need to complete them, and what tools you'll need to get the job done right. Pull on your grubby clothes and muck boots, because it's time to get dirty!

The Tools

Before we jump into cleaning, a quick rundown of the necessary tools will be helpful. If you have a well-stocked garage and garden shed, you likely have everything you need to care for the coop. If not, you can find most of these items at your local farm-supply store, hardware store, or garden center, or even online. I like to keep our supplies together in the shed so they are quick and easy to access. Here's what you need to have on hand and why.

au naturel

Some chicken keepers use rubber gloves and dust masks to prevent being exposed to or breathing the bird dander and airborne excrement. Immuno-compromised people and pregnant women, in particular, should protect themselves when cleaning the coop.

Wheelbarrow

Every chicken keeper needs a wheelbarrow or hand cart. These indispensible helpers make hauling supplies and waste easy and convenient, particularly when you're disposing of soiled bedding and manure. Made from materials such as hard plastic and long-lasting steel, they come in a range of sizes and styles to suit everyone. Choose one that matches your needs and physical ability.

- **PRICE RANGE:** $35 to $300
- **USE:** Hauling

Bucket

Another must-have item, a bucket can be used for myriad chores, such as mixing cleaners, refilling nests with bedding, and transporting all of your hand tools. You can find plastic buckets, aluminum buckets, buckets with lids, buckets with comfort-grip handles, and more. Select the one that's easiest for you to work with in both size and grip.

- **PRICE RANGE:** $2 to more than $20
- **USE:** Hauling, mixing

Scrapers

When it comes to scraping tools, just about anything goes—as long as it removes those concretelike chicken droppings. We've had the most success with tools from the hardware store's paint department, particularly a seven-in-one tool with a flat edge for scraping, a point for getting into tiny cracks, and a rounded edge that's perfect for removing droppings from the roosts. A wider putty knife, also from the paint department, works well for both scraping and scooping.

- **PRICE RANGE:** $1 to $10
- **USE:** Removing droppings

These scrapers work well in removing droppings, but you can use any tool that gets the job done.

Square-Point Shovel

Whether your coop has a dirt, concrete, wood, or mesh floor, you will need to remove the built-up chicken manure regularly—and that's where a square-point shovel and your wheelbarrow come in handy. The square point allows you to easily scoop heaps as well as scrape away any stuck-on debris. You can find aluminum, steel, and fiberglass shovels at your local hardware store.

- **PRICE RANGE:** $10 to more than $35
- **USE:** Scooping manure and bedding

Scrubbers

After scraping down the nest boxes, roosts, and walls and shoveling the manure, a thorough spring cleaning also includes scrubbing everything down using a scrub brush and cleaning product. You can use anything from a stiff-bristled hand-held brush to a plastic or metal scouring pad, depending on your coop and your preference. We use different-size brushes—a large brush with a comfort-grip handle, a traditional rectangular scrubber, and a tiny toothbrush (to get into all of the nooks and crannies).

- **PRICE RANGE:** $1 to $5
- **USE:** Removing droppings, sanitizing

Cleaning Products

You can find a range of bird-safe cleaners at your local pet- or farm-supply store, or you can make your own blend using items from your pantry. You will use the cleaner to sanitize the entire coop, including the food and water dishes. The commercial products use enzymes to effectively dissolve and remove droppings, but they can be pricey—especially if you have a larger coop. Homespun cleaners, such as a bleach-water solution (1 tablespoon bleach per gallon of water), a white vinegar-water solution (a 50:50 mix), and baking soda also do the trick for a lot less. For an easy cleaning solution you can make at home, check out the "Au Naturel" at left.

- **PRICE RANGE:** Pennies to more than $40
- **USE:** Cleaning, sanitizing

Pressure Washer

If your budget allows, purchase a pressure washer, which makes spring cleaning a breeze. This is a unit that shoots water (and cleaning agent, if desired) at high pressure, easily stripping droppings and stains from wooden boards and

au naturel

Baking soda is a real asset when it comes to cleaning. The following easy-to-make coop cleaner uses ingredients found in most home pantries. Baking soda acts as an abrader, and lemon juice contains natural grime-busting enzymes to break down organic matter. You'll need:

- A clean, empty spray bottle or bucket
- 3 cups of very hot water
- 2 tablespoons of baking soda
- 2 tablespoons pure lemon juice (freshly squeezed or bottled)

1. Pour the water into the bottle or bucket.
2. Add the baking soda, mixing it and making sure it completely dissolves and uniformly distributes throughout the water.
3. Add the lemon juice, again making sure you mix it well.

You can use this mixture just as you would any other cleaner. Though the baking soda and lemon juice are safe for your chickens, plan to thoroughly rinse and dry any cleaned surfaces. When you're done cleaning, dispose of the leftover solution.

We built our coop to be as convenient for us as it is for our chickens. Droppings fall below the coop for easy cleanup.

Manage Your Manure

In an enclosed chicken coop or a henhouse where chickens roost at night, the manure will collect—and fast. When cleaning and maintaining your coop, you'll have to decide how you'll manage the ladies' manure. Backyard chicken keepers have two options: an annual massive manure cleanup or smaller, more frequent cleanings.

If you choose to host a once-a-year coop-cleaning party and compost the droppings, you are essentially letting the manure and litter build up and begin its decomposing process (see the "*Au Naturel*" sidebar on page 29). Microbes and beneficial bacteria will go to work, creating heat and breaking down the material as the mixture ferments. Fly predators will swoop in and lay their eggs in the compost, helping to manage pesky flies. To manage the pile and prevent it from becoming too wet and reeking of ammonia, you simply add a layer of bedding (an inch or so to absorb any moisture) as needed. As the litter and manure decompose, their volume will naturally shrink.

If you choose frequent coop cleanouts—which could be a weekly or monthly chore, depending on the flock size, the henhouse, and your climate—you scoop the poop as it builds up, thus reducing the humidity and smell (and the yuck factor, if that's something that bothers you). The downside: Besides being somewhat labor-intensive and requiring more bedding, you also remove fly predators when you remove the manure. That could lead to a fly problem during the summer months, which you will have to manage.

- **WHEN:** Either once a year in the spring or more frequently throughout the year.
- **WHAT:** Plastic gloves, a shovel, a wheelbarrow, cleaning tools, and fresh bedding.
- **HOW:** For both annual and periodic cleaning methods, shovel all of the manure and bedding from the chicken coop into your wheelbarrow—making sure to get into all the nooks and crannies—and deposit it in your compost pile. Wash and sanitize the structure, allowing it to dry completely, and then lay down 4 to 6 inches of fresh bedding.

Annual Cleaning Chores

You're going to have to get dirty at least once a year. Like it or not, even a properly designed and maintained henhouse will need a *deep clean,* which involves moving furniture, scrubbing down walls and floors, and sanitizing everything. The amount of time the process takes will depend on the size of your coop, but here I've listed some guidelines for your annual thorough coop cleaning.

Sanitize the Furniture

Before you dive into bedding cleanup, first remove all of your coop's movable furniture and accessories and stow them where you can scrub them later. These pieces include your birds' feeding and watering bowls, supplement station, roosts, nesting boxes, and dust-bath box. They'll likely be soiled, stained, and in need of some good elbow grease.

- **WHAT:** A bucket, a hose, water, a scraper, a pot scrubber or piece of steel wool, and cleaning solution.
- **HOW:** Hose down surfaces and use a sharp-edged scraper to remove stuck-on droppings. Then, using steel wool or a plastic pot scrubber and some commercial or homemade cleaner (like a 50:50 vinegar-water solution), clean the accessories and inspect them to be sure they're in good working order. Rinse well, leave them in the sun to dry, and return them to the coop when your other chores are complete.

Remove the Bedding

Next comes the fun part: shoveling and removing the bedding. Whether you clean bedding out frequently or once a year, you'll need to remove it completely so that you can sanitize the structure's walls and floors.

- **WHAT:** Plastic gloves, a shovel or trowel, a broom and dustpan, and a wheelbarrow.
- **HOW:** Using your shovel or trowel, scoop all the bedding and manure from the chicken coop into the wheelbarrow and dispose of it appropriately (for example, in a compost pile). Make sure to get into all the nooks and crannies with your broom and dustpan.

This little garden trowel allows us to dig into the corners of the nest box to remove all soiled bedding and ensure a complete clean.

Scrape the Walls

As part of your annual deep-cleaning chores, you should scrape down and wipe off any dried-on manure from window ledges, the floor, nest boxes, and other flat surfaces where it's collected. When those droppings dry, they're like cement, so be prepared to use some muscle! Spraying them down with a high-pressure stream of water can help loosen stubborn spots.

- WHAT: Plastic gloves, a flashlight, scraping tools, steel wool, and a high-pressure hose or pressure washer (if needed).
- HOW: Scrape off dried manure as best you can. If necessary, use steel wool to rub out stuck-on droppings. In really stubborn cases, use water to dilute the manure and continue scraping with your flat edge or steel wool.

Sanitize the Coop

Finally, after you've scraped all of the coop's surfaces clean, it's time to sanitize. This annual routine, which involves cleaning with a bleach-water (1 tablespoon bleach per 1 gallon water) or vinegar-water solution (50:50), removes 95 percent of harmful pathogens, bacteria, and other disease-causing contamination that could threaten your birds. It also prevents pests, such as mites, from moving in.

- WHAT: Plastic gloves, a bucket, a hose, water, bleach- or vinegar-based cleaning solution, brushes, and sponges.
- HOW: Starting from the top and working your way down, wash each wall using your cleaning solution. Use your brushes and sponges to loosen any soiled spots you missed when scraping. Don't forget to thoroughly wipe the nest boxes clean. Rinse the surfaces well after sanitizing and allow the coop to dry completely.

Annual Maintenance Chores

It's a good idea to also do a thorough structural inspection of your chicken coop once a year, and a great time to do this is when you're doing your

You'll be surprised where manure ends up—this is where the various paint scrapers come in handy.

Bird Brain

What do you do with your ladies when you're cleaning their coop? If you have them in a fenced area or outdoor chicken run, they'll be content hunting and pecking for grubs while you scrub. If not, you can easily set up some temporary housing using a circle of chicken wire bound with zip-ties. Make sure there are no exposed sharp points or edges that could harm your birds, though.

The key to doing any chore is to make it a habit. The more often you clean, the cleaner your coop will keep.

annual deep cleaning. While you're scrubbing away, take a look at the building features and make any necessary repairs. As always, use care and common sense when using power tools and sharp objects.

To make sure your coop will weather whatever storm is on the horizon, visually inspect the structure for overall integrity. (This is similar to your daily yard check, just more thorough.) Look at the coop's base, legs, and ramps. Are they solid and secure? Check the building's interior and exterior for loose boards, protruding nails, and large draft holes, repairing them as necessary. Examine the coop's doors, including nest-box doors and hatches. Are the hinges screwed in tightly? Do your hooks and locks work properly?

> **SQUAWK BOX**
>
> ***"I can truly say that, during the time I have reared poultry, I have ever found the pursuit to be a 'labor of love,' and (like virtue) 'its own reward.' I feel that, though more tangible benefits had not fallen to my lot, I can still look back on the hours spent among my feathered pets with affectionate gratitude."***
>
> ***Mrs. Elrington Douglas Arbuthnott***

You should also look for signs of predators and shore up any varmint protection you have. If you see rat or mole holes around your henhouse, put some hardware cloth across the opening and cover it with dirt. If you're getting frequent fly-bys from area raptors, consider deterring them with some high-strung wire adorned with reflective tape or old compact discs.

Another important chore to check off once a year: Make certain your coop's roof is sound and watertight. If your roof has shingles, check for missing tiles and replace them. If you have a plywood or metal roof, check that it's securely nailed down. Clean out gutters if need be. And if you have a tarp covering your coop or run, inspect it for holes and lay down a new one if necessary.

If you have plumbing or electricity running to your chicken coop, inspect that annually as well. Make sure you repair leaks in your pipes or fittings as needed. Look at your wires and electrical connections and make sure everything is safe and working properly.

- **WHAT:** A flashlight, along with any tools and hardware you might need.
- **HOW:** Walk around the henhouse and chicken yard, making sure the structures are sound, the fences are solid, the roof is secure, the yard is predator-proof, and the buildings are in good working order.

Refeather the Nest

Coop maintenance isn't difficult, but it's something that needs to be done. With a little dedicated time, you can keep the henhouse clean and orderly. This benefits the birds (who wants to live in a dirty house, right?) as well as the humans (a clean coop is a pleasure to be around).

Start with a sound roof on your coop and check it annually for any wear.

Trouble-Shooting

When keeping a chicken coop, problems may arise that you're just not sure how to address. You may notice a hole by the henhouse that's been getting bigger and bigger each night. Eggs go missing or their shells are pecked. The ladies start scratching themselves and seem to have little black nits on their feathers. What's going on? Oh—and what do I do when it starts snowing?

After keeping chickens for years now, we've had some complicated chicken-coop puzzles to figure out. One that I'll never forget was determining what kind of critter snuck into the chicken coop and killed two hens and Larry, our best rooster. The night the attack occurred, we arrived home after dark and headed out to the coop to put the chickens to bed. That's when we saw the carnage: Larry's remains were at the scene, but two of the girls were missing. The next morning, we walked the property and discovered one of the hens toward the bottom of the hill, but one was still unaccounted for. We figured something big had dragged her away. But later, she turned up (or *was* turned up) in the chicken yard. The marauder had buried the hen in a pile of fallen leaves against the fence, saving her for later.

The villain turned out to be a bobcat. Bobcats are notorious for burying their kills and returning for a second course. We handled the situation by being extra vigilant about locking the ladies in before sundown (so much for nights out on the town!).

Aside from structural problems that can be fixed with a drill and some screws, other puzzles will inevitably surface. I've categorized them into three groups:

- Pests and predators
- Seasonal management
- Chicken behavior

Managing Pests and Predators

Identifying and eradicating pests and predators that invade a chicken coop can be a challenge. Pests include insects and mites; predators include four-legged and flying dangers such as raccoons and hawks. In each case, the first thing to do is figure out what kind of intruder you're dealing with. Below, read about the signs and solutions for a number of pests and predators.

Small rodents can gain entry to your coop through holes smaller than the size of a quarter.

Invasion of the Mites and Lice

External parasites, such as lice and several species of mite, like the taste of chicken. Mites eat the chicken's skin, feathers, and blood, while lice chew on and irritate the bird, causing her to break off or pull out her feathers in frustration. These

Dust baths might not sound very clean to you, but they help your chickens stay free of itchy, bothersome bugs.

While grazing can be endlessly beneficial to your girls, it can also become a hazard to those with weak immune systems.

types of parasites are ubiquitous, but good henhouse management can keep them at bay—as can insect and mite repellents, such as diatomaceous earth, sprinkled in the hens' dust-bath boxes.

- SIGNS AND SYMPTOMS: Mites will appear as red or black specks crawling on the birds' skin, around their vents, on their eggs, or in their nests; the birds may have pale combs and wattles as well as suffer weight loss and a drop off in egg laying. Lice appear as pale insects on the skin, and their eggs appear as white clumps at the base of the birds' feathers; the chickens' plumage may look dull or rough.
- HOW TO HANDLE IT: When your coop has been infested, the best thing to do is dust every inch of the dwelling and the birds with poultry-approved insecticide from your local farm-supply store. Follow the instructions to a T, and keep the coop clean.

Worms, Protozoa, and Other Delicacies

If you keep your hens on the same patch of ground year after year, internal parasites such as roundworms, flatworms, and the protozoa *Coccidia* (which can cause coccidiosis, the most common cause of death in young birds) can become a problem. It makes sense: As the chickens peck the ground, they pick up all kinds of interesting bugs, including internal parasites. With proper management and hygiene, most chickens develop a resistance to parasitic worms.

- SIGNS AND SYMPTOMS: In general, birds with worms will appear weak and listless, will lose weight, and may even display respiratory distress and behavior changes. Young birds with *Coccidia* may have loose, watery, or bloody droppings; older birds will decrease their egg laying.
- HOW TO HANDLE IT: Before you medicate your flock, send some stool samples to your veterinarian and develop a deworming schedule based on your birds' needs. Also step up your coop-cleaning regimen.

Flies, So Many Flies

Because flies thrive in damp litter and manure, these little pests may become a problem in your chicken coop. Besides being incredibly irritating, they can also spread disease and transmit tapeworms to those ladies who like to eat flies and their larvae. Chickens will help control the little buggers, but you can help them by keeping the coop free of standing water and overly wet bedding.

- **SIGNS AND SYMPTOMS:** Flies. Everywhere. Enough said.
- **HOW TO HANDLE IT:** In addition to keeping the litter dry by frequently adding fresh bedding, set out fly traps or fly paper and introduce natural fly predators (including your chickens!). If possible, avoid using insecticides.

Wild Birds

Where there's bird food, there are birds—including wild birds. If your chicken coop has open access to the sky, all kinds of birds will stop by to visit, particularly if you feed whole-grain scratch to your ladies as a treat. Though it's fun to watch the robins and sparrows play with your chickens, these visitors can spread parasites and disease.

- **SIGNS AND SYMPTOMS:** You'll see birds swoop in and steal your chickens' food, or you may notice the feed isn't lasting as long as it once did. Wild birds can be seasonal, so you may see them visit more often during certain times of year.
- **HOW TO HANDLE IT:** If it's a real problem, you might consider building an enclosed pen. Other options include stringing wire or fishing line across the yard, which stops the birds from flying in, and hanging reflective tape or old compact discs, which reflect the sun and momentarily blind the birds. You can also keep them out of the food simply by putting the feed troughs inside the henhouse.

Your girls aren't the only birds that occupy your yard. Make sure you account for the wild ones that might be invading your ladies' trough.

You don't want a fox in the henhouse! Predator prevention is the only way to keep your chickens safe.

Predators on the Ground and in the Sky

Ground predators, such as raccoons, opossums, coyotes, skunks, cats, dogs, minks, and even foxes, pose a real threat to chickens. Flying predators, such as hawks and owls, are a problem too. When one strikes, it's heartbreaking. All you want to do is figure out what did that to your bird and how to stop it from happening again. Different predators leave different calling cards, including tracks, method of maiming or killing, what they take, and how they take it. I recommend that all chicken owners invest in a good book that describes predator problems in detail. You should know which predators exist in your area, learn their habits, and design your coop with those bad guys in mind.

- SIGNS AND SYMPTOMS: Depending on the predator, you may see tracks, broken eggs or eggshells around the coop, missing eggs or hens, maimed or dead hens, or feathers but no bodies.
- HOW TO HANDLE IT: Determine what kind of predator you're dealing with. For the larger animals, you may need a live trap or the help of an expert tracker. For the smaller predators, you could also use a trap or install electric fencing, which zaps and scares off predators while the hens sleep. If you have a problem with flying predators, string wire above the coop. When they fly down into the yard, they'll get hung up on the wire and not be too keen on returning. If you have a large flock, consider a guard animal, such as a reliable dog or llama.

If your dog is chicken-friendly, enlist him for the role of chicken protector.

In general, however, you should make sure your coop has these predator-proof features:

- **Adequate fencing.** As we discussed previously, fencing matters. You want to choose a type of fencing that's strong and has a tightly woven mesh to keep diggers and strong-handed varmints out. If you have a completely enclosed coop, make sure the wire is strong and secure.
- **Adequate height and depth.** The fencing should be high enough to keep creeping predators out and buried low enough into the ground to ward off digging predators. If you see any signs of diggers, check the fencing and make sure it's in place and not corroded or pushed out of the way.
- **Secure hardware.** You'd be amazed at what raccoons can do with their little hands. They can figure out how to open latches, open doors, rip through mesh—all to get to the tasty chickens inside the henhouse. Make sure that your latches work properly and are out of reach of tiny fingers. Regularly inspect vent covers as well to be sure there are no holes or entry points.

Seasonal Management

Whether you live in an area that's temperate year-round or one that experiences all four seasons' climate extremes, you can expect some temperature- and weather-related challenges to contend with. The extent to which you and your ladies deal with seasonal challenges will, obviously, depend on where you live. But here are some general situations and how to manage them in your chicken coop.

Chilly Chickens

When temperatures dip in the wintertime, your birds will feel the chill. Though their plumage, physical activity, metabolism, and huddling behavior will keep them toasty warm, they can fall victim to frostbite if the mercury plunges too low—particularly in a damp coop. Humidity in the air can chill the chickens by drawing their body heat to the surface to vaporize. If temperatures are low enough to freeze moisture in the air, they can freeze moisture on your birds' combs and wattles.

- SIGNS AND SYMPTOMS: Frozen combs and wattles appear pale, and seriously frozen ones shrivel and may eventually die back. Frozen combs and wattles that have thawed appear red, hot, swollen, and painful; the bird most likely will have no interest in moving or eating.
- HOW TO HANDLE IT: Prevention through proper management is the best way to handle chilly chickens. Make sure your henhouse is well ventilated to remove moist air from the space, and place your

perches in the least-drafty areas of the coop. If you live in a particularly cold region, install a heat lamp plugged into a thermostat to keep temperatures above freezing inside the henhouse. And to keep their metabolisms firing, feed the ladies some extra scratch in the morning and night; stimulate their appetites by offering some mash moistened with warm milk or water.

Stinky Digs

As I advised in chapter 7, coop maintenance is critical to keeping your hens (and your neighbors!) happy. One sure sign that your henhouse needs some spring cleaning is if it starts, well, getting a little stinky. While being cooped up for a long winter slumber, all the ammonia and off-gases from your girls' droppings can build up. And when the warm spring sunshine cooks the mixture, the result is a ripe henhouse.

- **SIGNS AND SYMPTOMS:** Wafts of odor emanating from your henhouse (possibly accompanied by complaints from neighbors).
- **HOW TO HANDLE IT:** Flip back to chapter 7 for instructions on how to give your henhouse a thorough cleaning. If you've already done your annual scrub-down, you may need to toss some fresh bedding on top of the old. This will absorb moisture and cut down the odor.

Fried Chickens

In the sunny summertime, chickens run the risk of overheating. When temperatures reach 104 degrees Fahrenheit or above, the birds can't lose excess heat fast enough to maintain their body temperature.

- **SIGNS AND SYMPTOMS:** Your temperature gauge reads more than 100 degrees Fahrenheit and you see your girls scurrying to shady spots, sitting in place and panting with their beaks open, standing with their wings away from their bodies, or drinking excessive amounts of water. These birds are hot, and they need some relief.
- **HOW TO HANDLE IT:** As when managing your birds in cold-temperature extremes, prevention is key. When temperatures climb, increase the number of watering stations in your chicken coop, fill them frequently with cool fresh water, and place them in the shade to keep the water cool and palatable. Also open the doors and windows to your henhouse, increasing airflow as much as possible, and provide lots of shady spots, such as lean-to structures and A-frames, where the chickens can rest.

Slowed Egg Production

In the fall when the hours of daylight dwindle, you may notice your ladies' egg production decrease. Don't worry—this natural slow-down occurs when the hours of sunlight fall below fourteen (and when your hens are getting on in age, but that's another topic altogether!).

- **SIGNS AND SYMPTOMS:** You'll know when your hens' egg yield drops from their daily lay to one egg every few days.
- **HOW TO HANDLE IT:** When this happens, you have two choices: you can let the ladies rest for the season, or you can install a full-spectrum lamp on a timer that mimics sunlight and extends the "daylight" to the required fourteen hours.

Perches help chickens work out their hierarchy on their own, which means fewer fights. Also provide space and more feeding stations.

Fowl Behavior

When something's amiss with the henhouse, the hens may start to act up and display some odd behavior, such as harassing one another or pecking at each other's eggs. What to do? Though some behaviors may be related to the chickens' health and should be evaluated by a veterinarian, some can be managed with coop solutions.

Henpecking

A normal chicken flock has a pecking order, which means the hens (and roosters) have a social hierarchy. Those higher in the pecking order have first dibs on things such as food and roosting spots. When someone lower in the order challenges those higher-ups, spats may occur. The birds may fight, pull out each other's feathers, peck at each other, and scratch each other with their sharp claws. To a point, this is normal behavior, but the fights can be brutal, especially when an entire flock targets one bird. This sometimes happens when new birds are introduced to a flock.

- SIGNS AND SYMPTOMS: If one of your hens has sores on her head or body, has lost feathers in spots, seems bullied by the other birds, has lost weight, or looks otherwise ill, she may be the target of henpecking.

SQUAWKBOX

"If you have a flock of birds, make sure to provide enough feeders such that the local bully can't prevent the other birds from eating."

Hilary Stern, DVM

- **WHAT TO DO:** First of all, separate the henpecked bird from the others and give her a chance to heal. Then, give the chickens more space (if possible) to encourage a stable pecking order and so that the lower ranking birds have more space to get away from the higher-ups. You should also make sure your coop has enough hiding places, feeders, and waterers to accommodate the flock.

A Taste for Eggs

Egg pecking and egg eating can be an addicting habit for chickens. Once a bird develops a taste for eggs, it's almost impossible to get her to not put a hole in each egg in the nest—which means you have fewer eggs to eat. The habit starts when a chicken tries out an egg that broke in the nest, which was likely caused by poor management, overcrowding, or not enough nesting boxes.

- **SIGNS AND SYMPTOMS:** You'll know the broken eggs when you see them, but it will be hard to tell which bird is the perpetrator. If you suspect a hen is breaking the eggs and eating them, try to catch her in the act. Otherwise, look at the ladies' beaks and check them for dried yolk.
- **WHAT TO DO:** This is one problem you really have to stop before it happens. Gather eggs several times a day and make sure you have enough nesting boxes for your laying hens. Unfortunately, if a chicken starts eating eggs, you may have to separate her from the flock.

Check your nest boxes often for eggs to keep them from getting dirty or being crushed.

Take It as It Comes

When designed and managed properly, chicken coops make ideal homes for chickens. Once in a while, though, your birds may encounter trouble that affects their health and well-being. It's up to you as a responsible chicken keeper to be aware of and prepared for problems before they start.

Glossary

avian: Having to do with birds; avian medicine, avian art, etc.

aviary: A confined area for keeping birds.

bantam: A miniature chicken weighing 1 to 3 pounds and one quarter to three-eighths the size of large breed chickens; some are scaled-down versions of large breeds but a few bantam breeds have no full-size counterparts; slang: banty (plural: banties).

bedding: The thick layer of cushiony absorbent material used to line nesting boxes and poultry-building floors; also called *litter*.

broiler: A tender nine- to sixteen-week-old meat chicken generally weighing 2½ to 3½ pounds; also called a *fryer*.

brooder: A small enclosure for peeps used to imitate the warmth and protection they would receive from their mothers.

broody hen: One who, through hormonal changes, stops laying and elects to hatch eggs or care for baby chicks; also called a *broody*.

cannibalism: The habit stressed chickens have of pecking other chickens or themselves until they draw blood; a bloodied chicken often is killed by her peers.

coccidiostat: A drug used to prevent the common protozoal infection, *coccidiosis*; often added to

commercial chick starter rations.

comb: The fleshy red appendage atop a chicken's head.

coop: The pen or building where chickens reside.

deep bedding system: A coop-cleaning system where litter is continually added until the coop is cleaned once a year.

dust bathing: The act of a chicken wallowing in dirt to clean its feathers and discourage external parasites.

flock: A group of chickens.

free-range chickens: Uncaged fowl allowed to forage wherever they choose.

fryer: A tender nine- to sixteen-week-old chicken generally weighing 2½ to 3½ pounds; also called a *broiler.*

Gallus domesticus: The latin name for the domestic chicken.

gizzard: The tough internal organ where food is macerated.

grit: Pebbles, sand, or a commercial "grit" product ingested by a chicken and used by the gizzard to grind food.

hen: A female chicken at least one year old.

heritage chicken: An antique breed of chicken that has been purposefully preserved through the efforts and to the standards of the American Livestock Breed conservancy.

humidity: The amount of moisture in the air.

layer, laying hen: A chicken kept for egg production.

litter: The thick layer of cushiony absorbent material used to line nesting boxes and poultry-building floors; also called *bedding.*

nest: A dark, secluded place where a hen feels it's safe to lay her eggs.

nesting box: A man-made cubicle placed in henhouses so that hens can lay their eggs away from the main flow of traffic; also called a *nest box.*

pastured poultry: Fowl raised in a pasture setting and housed in movable shelters.

pecking order: The social order of chickens.

perch: (noun) The place, usually elevated rails, where chickens sleep at night; also called a *roost.* (verb) The act of resting on a roost.

pop hole: A chicken-size door on a coop.

pullet: A female chicken less than one year of age.

roost: (noun) The place, usually elevated rails, where chickens sleep at night; also called a *perch.* (verb) The act of perching.

scratch: (verb) The act of scratching the ground in search of food. (noun) Any grain product fed to chickens.

set: The act of allowing or encouraging a broody hen to incubate eggs.

starter: Commercial feed ration formulated for newly hatched chicks; there are two formulations: regular and broiler.

wattles: Two dangles of red flesh drooping down from the outer edges of a chicken's chin.

zoning: Laws regulating land use, including whether or not landowners can keep chickens.

Resources

Retailers

Absorbent Products Ltd.
www.absorbentproductsltd.com
Manufactures an environmentally friendly bedding material.

Animal Supplies International
www.animalsupplyintl.com
Sells agricultural supplies, including coop furnishings.

C & S Products
www.cs-prod.com
Sells feed for wild birds, which can help keep them away from your girls' lunch.

Cargill Animal Nutrition
www.cargill.com/feed
Supplies feed and nutritional information.

Cashton Farm Supply
www.cfspecial.com
Manufactuers and distributes organic feed as well as poultry equipment.

Chicken Doors
http://chickendoors.com
Manufactures an automatic henhouse door that opens in the morning and shuts in the evening.

Chickens for Backyards
www.chickensforbackyards.com
Sells peeps, feed, and supplies.

Countryside Organics
www.countrysideorganics.com
Manufactures organic poultry feed.

Creative Coops
www.creativecoops.com
Manufactures a line of build-your-own-coop kits.

Cutler's Pheasant and Poultry Supply
www.cutlersupply.com
You can find much of what your need for your chickens on this site: hawk netting, galvanized feeders, wire, and more.

Easy-Garden
www.easy-garden.com
Sells gardening and poultry-keeping products, including supplies for vermicomposting.

Eggboxes
800-326-6667
http://eggboxes.com
Sells an array of hatchery and egg-seller supplies, including many types of plain or custom-printed egg cartons.

Eggcartons.com
888-852-5340
www.eggcartons.com
Visit this site to order plain or custom-imprinted paper, foam, and plastic egg cartons at discount prices. Also available are incubators, feeders, waterers, books, and chicken-themed gifts.

Egg Cartons Online
877-454-3447 (EGGS)
www.eggcartonsonline.com
Markets a huge line of egg cartons, an egg washing kit, hatching supplies, and books.

Estes Hatchery
www.esteshatchery.com
Sells peeps, chicken-keeping supplies, furnishings, and medications.

Fleming Outdoors
www.flemingoutdoors.com
A comprehensive online farm-supply store.

Formex Manufacturing, Inc.
www.formex.com
Manufactures an easy-to-put-together prefabricated chicken coop.

Fowl Stuff
www.fowlstuff.com
A plastic nest box solution.

GQF Manufacturing
www.gqfmfg.com
Sells everything you need to breed chickens and house and care for peeps.

Grandpa's Feeders
www.grandpasfeeders.com
Manufactures a unique chicken feeder that protects feed from weather and pests.

Happy Hen Treats
www.happyhentreats.com
Manufactures a line of treats for chickens as well as products such as a treat ball.

Harris Farms, LLC
www.harrisfarmsllc.com
Manufactures a line of feeders and waterers.

Hoover's Hatchery, Inc.
http://hoovershatchery.com
Sells a variety of peeps as well as chicken-keeping supplies.

IPS Care Free Enzymes
www.carefreeenzymes.com
Manufactures environmentally and chicken-friendly enzyme products such as egg wipes and odor eliminators.

Kemp's Koops
888-901-2473
www.poultrysupply.com
Kemp's Koops sells chicken, waterfowl, gamebird, and exotic-avian supplies and incubators.

Kencove Farm Fence Supplies
www.kencove.com
Manufactures and supplies fencing products.

Kent Feeds
www.kentfeeds.com
Manufactures poultry feeds.

MannaPro
www.mannapro.com
Manufactures a variety of solutions for the coop, including odor neutralizers and water protectors.

Meyer Hatchery
www.meyerhatchery.com
Sells peeps and offers a comprehensive catalog of poultry-keeping products.

Modesto Milling
www.modestomilling.com
Manufactures certified organic poultry feed.

Murray's Hen Hoops
www.henhoops.com
Manufactures a variety of mobile chicken coops.

My Pet Chicken
www.mypetchicken.com
Another comprehensive site that sells everything your chickens need in a coop.

Nite-Guard Solar
www.niteguard.com
Manufactures a solar-powered weatherproof light that flashes all night to ward off predators.

Omega Fields
www.omegafields.com
Manufactures feed supplements, including one for laying hens.

Omlet
www.omlet.us
The American website of this British prefab-coop manufacturer features helpful information on raising chickens.

The Poultry Butler
www.poultrybutler.com
Manufactures an automatic door for henhouses that opens and closes using either a light sensor or a timer that you set.

Poultryman's Supply Company
812-603-7722
http://poultrymansupply.com
Poultryman's Supply offers incubators, brooders, waterers and feeders, medications, books, leg bands, egg cartons, and much, much more.

Premier 1 Supplies
www.premier1supplies.com
This site sells everything you need—from fencing to feeders—to set up your coop.

Quail Manufacturing
www.eggcartn.com
Manufactures chicken tractors in two sizes.

Randall Burkey Company
www.randallburkey.com
Sells peeps as well as a bevy of chicken-keeping supplies.

Shop the Coop
www.shopthecoop.com
A large online supplier of all things chicken.

Smith Poultry & Game Bird Supply
www.poultrysupplies.com
Sells incubation supplies, brooders, nesting boxes, netting, leg bands, waterers, feeders, medications, vaccines, vitamins, disinfectants, books, and a lot of other items.

Stromberg's Chicks & Gamebirds Unlimited
www.strombergschickens.com
Sells a variety of peeps and poultry-keeping supplies.

The Sweeter Heater
www.sweeterheaters.com
A safe, energy-efficient infrared heat source for pets.

Tractor Supply Co.
www.tractorsupply.com
Sells a variety of chicken-keeping supplies and furnishings, as well as several prefabricated coops.

Treats for Chickens
www.treatsforchickens.com
Much more than treats, this company also sells supplies, toys, furnishings, treatments, and cleaners.

Welp Hatchery
www.welphatchery.com
Sells peeps as well as brooder supplies.

Organizations

American Bantam Association (ABA)
www.bantamclub.com
Founded in 1914, the American Bantam Association promotes the breeding, exhibition, and selling of purebred bantam chickens and ducks. Visit its website to join the ABA, access breed-club and member links, purchase books and other merchandise, and read about the organization's many programs.

American Livestock Breeds Conservancy
www.albc-usa.org
The American Livestock Breeds Conservancy works to protect nearly one hundred breeds of cattle, goats, horses, asses, sheep, swine, and poultry from extinction. Clink on "Heritage Chicken" to access a treasure trove of information about old-time, endangered breeds.

American Pastured Poultry Producers Association (APPPA)
http://apppa.org
The APPPA unites pastured poultry producers and distributes pastured poultry resources to consumers and potential producers. Visit their website to download the APPPA brochure or locate a pastured poultry producer in your locale.

American Poultry Association (APA)
www.amerpoultryassn.com
Founded in 1873, the American Poultry Association sanctions poultry shows and publishes the APA Standard of Perfection (the rules by which

purebred poultry is shown), a yearbook, and a quarterly newsletter. Use the pull-down menus at the APA website to access a variety of avian information; their Health series and "Raising Birds in the City" (find it in the "useful information" menu) are especially well written.

Rare Breeds Survival Trust (RBST)
www.rbst.org.uk
The Rare Breeds Survival Trust (the United Kingdom's equivalent of the American Livestock Breeds Conservancy) currently montors thirty-one breeds of rare chickens, including many concurrently tracked in North America by the ALBC.

Society for the Preservation of Poultry Antiquities (SPPA)
www.feathersite.com/Poultry/SPPA/SPPA.html
Fascinated by heirloom chickens? Join the SPPA and help preserve and promote them.

Suggested Reading

The following books and publications are geared toward small- and medium-scale chicken keepers.

Books

Damerow, Gail. *Storey's Guide to Rasing Chickens: Care, Feeding, Facilities*. North Adams, MA: Storey, 2010. Gail Damerow knows her chickens. Whether you keep laying hens or raise chickens for meat, and no matter your level of expertise, this is a book you'll refer to time and time again.

Litt, Robert and Hannah. *A Chicken in Every Yard: The Urban Farm Store's Guide to Chicken Keeping.* New York, NY. Ten Speed Press, 2011. This informative guide to chicken keeping has sustainability and community in mind.

Weaver, Sue. *Hobby Farms® Chickens, Second Edition*. Freehold, NJ: Hobby Farm Press®, 2010. Sue Weaver offers sound advice based on years of adventures in poultry with her husband, John. This updated edition provides the information you need to succeed with chickens on your hobby farm or in your home and yard.

Magazines

Backyard Poultry
www.backyardpoultrymag.com
Visit this bimonthly publication's Web pages to read scores of articles archived on-site.

Chickens
www.hobbyfarms.com/chickens-magazine
Chickens is published six times a year by the folks who bring you *Hobby Farms, Hobby Farm Home,* and *Urban Farm*. Check it out!

Hobby Farms
www.hobbyfarms.com
Hobby Farms is *the* magazine for rural enthusiasts—hobby farmers, small production farmers, and those passionate about the country. This bimonthly magazine is devoted to all aspects of rural life, from small farm equipment, to livestock, to crops.

Hobby Farm Home
www.hobbyfarms.com/hobby-farm-home-portal.aspx
Hobby Farm Home is the home magazine for those truly living in the country. It highlights farmhouse activities such as cooking, crafting, collecting, pet care, and home arts and skills.

Poultry Press
www.poultrypress.com
Poultry Press, established in 1914, is an information-packed monthly not to be missed.

Poultry World
From the folk at I-5 Publishing, this bimonthly magazine focuses on chickens and their fascinating cousins.

Urban Farm
www.urbanfarmonline.com
From the editors of *Hobby Farms* and *Hobby Farm Home*, *Urban Farm* magazine reaches out to those in the city and suburbs who want to raise chickens, grow food for themselves, support local agriculture, and live more sustainably.

Websites

An All-Breed Chart
www.ithaca.edu/staff/jhenderson/chooks/ch-links.html
John. R. Henderson, a librarian at Ithaca College and a Lodi, New York, hobby farmer, has created the ultimate poultry-link website. While you're there, click on the "ICYouSee Handy Dandy Chicken Chart" link.

Association of Avian Veterinarians
www.aav.org
When your sick or injured chicken needs a specialist, find one via the Association of Avian Veterinarians website.

Backyard Chickens
www.backyardchickens.com
This great site features articles, coop plans, a forum, and more.

Brown Egg, Blue Egg
www.browneggblueegg.com
Brown Egg, Blue Egg belongs to Alan Stanford, PhD. If you love chickens (click on "Stories") or want to learn more about them (the selection of articles at this site is beyond extensive), don't miss Brown Egg, Blue Egg.

Chicken Crossing
http://chickencrossing.org
A great site dedicated to city- and pet-chicken keepers that includes a good message board.

Chicken Breeds
www.ansi.okstate.edu/breeds/poultry/index.htm
Click on "Chickens" to read about and view photos of any breed you can think of and then some.

Chicken Feed: The World of Chickens
www.lionsgrip.com/chickens.html
The Chicken Feed website brings you "sources of natural chicken feed, knowledge about traditional ways of feeding chickens around the world and in old times, and health before profit in raising and feeding chickens."

Duluth City Chickens
http://duluthcitychickens.org
This site is about city chickens in the far North and has a good FAQ section.

FeatherSite
www.feathersite.com
Don't miss FeatherSite—it's amazing! Links lead to every conceivable poultry topic, organization, and business online.

Mad City Chickens
www.madcitychickens.com
Chicken keepers in Madison, Wisconsin, bring you a coop preview and a very helpful FAQ section.

Mad City Chickens on YouTube
www.youtube.com/group/madcitychickens
The Mad City chicken keepers bring you more than 600 great chicken videos.

Merck Veterinary Manual
www.merckvetmanual.com
The online version of the Merck Veterinary Manual encompasses more than 12,000 topics and 1,000 illustrations searchable by topic, species, specialty, disease, and keyword. Access is free, compliments of Merck, Inc.

The City Chicken
http://thecitychicken.com
This comprehensive site features a chick tractor gallery, great articles, FAQs, oodles of great pictures, and the best roundup of chicken laws online.

The Coop
www.the-coop.org
Looking for a truly comprehensive resource for chicken keepers? Here it is!

Urban Chickens
http://urbanchickens.org
Urban Chickens brings you a blog, a forum, articles, and information on the legalities of keeping chickens in town.

Urban Chickens Network Legal
Resource Center
http://wiki.urbanchickens.net
The Urban Chickens Network Legal Resource Center brings you a list of cities where you can legally keep chickens along with each city's specific statutes.

USDA Cooperative Extension Service Finder
http://www.csrees.usda.gov/Extension/
This site will help you find you nearest cooperative extension service.

Index

A

I

L

M

About the Author

Award-winning writer Wendy Bedwell-Wilson writes about pets from her 80-acre hobby farm in southwest Oregon, which she shares with her husband, a retired racing Greyhound named Magic, a hound mix called Pete, domestic shorthair cats Bubba and Benny, and a menagerie of barnyard critters, including resident roosters Larry and Lance and their egg-laying ladies. She regularly contributes to more than a dozen magazines, including *Chickens*, *Dog Fancy*, *Dog World* and *Cat Fancy*, and has had more than 650 feature stories published to date. She has also written six dog breed books as well as a continuing education text book for Animal Behavior College about how to launch and operate a pet-sitting/dog-walking business. Wendy's passion centers on educating people through well-crafted prose about how to better bond with their animals and live healthier lives together.